Wireless Networks

Series editor

Xuemin Sherman Shen
University of Waterloo, Waterloo, Ontario, Canada

More information about this series at http://www.springer.com/series/14180

Liang Xiao • Weihua Zhuang • Sheng Zhou
Cailian Chen

Learning-based VANET Communication and Security Techniques

 Springer

Liang Xiao
Department of Communication Engineering
Xiamen University
Xiamen
Fujian, China

Weihua Zhuang
Department of Electrical & Computer
Engineering
University of Waterloo
Waterloo
ON, Canada

Sheng Zhou
Department of Electronic Engineering
Tsinghua University
Beijing, China

Cailian Chen
Department of Automation
Shanghai Jiao Tong University
Shanghai, China

ISSN 2366-1186   ISSN 2366-1445   (electronic)
Wireless Networks
ISBN 978-3-030-13192-0   ISBN 978-3-030-01731-6   (eBook)
https://doi.org/10.1007/978-3-030-01731-6

This Springer imprint is published by the registered company Springer Nature Switzerland AG
The registered company address is: Gewerbestrasse 11, 6330 Cham, Switzerland

Preface

This book provides a broad coverage of the vehicular ad hoc networks (VANETs) to support vehicle-to-vehicle communications and vehicle-to-infrastructure communications and focuses on the vehicular edge computing, vehicular network selection, VANET authentication, and jamming resistance. Machine learning-based methods are applied to solve these issues. This book includes 6 rigorously refereed chapters from prominent international researchers working in this subject area. Professionals and researchers will find *Learning-Based VANET Communication and Security Techniques* a useful reference. Graduate students seeking solutions to VANET communication and security-related issues will also find this book a useful study guide.

In Chap. 1, we briefly introduce the vehicular communication and VANET security and review the machine learning techniques that can be applied in VANETs.

In Chap. 2, we discuss the VANET authentication and focus on the reinforcement learning-based rogue edge detection with ambient radio signals.

In Chap. 3, we review a multi-armed bandit-based offloading scheme for vehicular edge computing.

In Chap. 4, we present an intelligent network selection scheme to provide real-time services for vehicular systems.

In Chap. 5, we review the UAV-aided vehicular transmissions against jamming and formulate a stochastic game between the UAV and the jammer. A reinforcement learning-based UAV relay scheme is presented, and its performance is evaluated.

In Chap. 6, we conclude this book with a summary and point out several promising research topics in the learning-based VANET communication and security techniques.

This book could not have been made possible without the contributions by the following people: Xiaozhen Lu, Geyi Sheng, Minghui Min, Dongjin Xu, Xiaoyue Wan, Xingyu Xiao, Yuliang Tang, Yuxuan Sun, Xueying Guo, Jinliui Song, Zhiyuan

Jiang, Zhisheng Niu, Xin Liu, Shumin Bi, Yuying Hu, and Tom H. Luan. We would also like to thank all the colleagues whose work enlightened our thoughts and research made this book possible.

Xiamen, China   Liang Xiao
Waterloo, ON, Canada   Weihua Zhuang
Beijing, China   Sheng Zhou
Shanghai, China   Cailian Chen
August 2018

Contents

Chapter 1
Introduction

Vehicular Ad Hoc Networks (VANETs) provides the efficient dissemination of information among the vehicles and roadside infrastructure. However, due to the high mobility of onboard units (OBUs) and the large-scale network topology, VANETs are vulnerable to attacks. In this chapter, we first review the fundamentals of VANETs in Sect. 1.1. Next, the type of attacks in VANETs is presented in Sect. 1.2, including the scope of the attack, and the impact to VANETs. We review the VANETs security solutions based on machine learning techniques including supervised learning, unsupervised learning and reinforcement learning in Sect. 1.3. Finally, we conclude in Sect. 1.4.

In this book, we briefly introduce communication and security in VANETs and investigate the techniques based on machine learning to improve communication efficiency and anti-jamming performance in Chap. 1. We propose a physical-layer rogue edge detection scheme based on reinforcement learning for VANET in Chap. 2. In Chap. 3, we establish a learning-based task offloading framework based on the multi-armed bandit. In Chap. 4, we apply a fuzzy-based method to make network selection on the heterogeneous vehicle network. We investigate anti-jamming solutions with the aid of UAV in VANETs and propose the UAV relay against smart jamming with reinforcement learning in Chap. 5. In Chap. 6, we conclude this book with a summary and point out several promising research topics in communication and security in VANETs.

1.1 Vehicular Communications

With mobile operating systems becoming increasingly common in vehicles, it is undoubted that vehicular demands for real-time Internet access would get a surge in the soon future [1–8]. The VANET offloading represents a promising solution to the overwhelming traffic problem engrossed to cellular networks. The wide

© Springer Nature Switzerland AG 2019
L. Xiao et al., *Learning-based VANET Communication and Security Techniques*, Wireless Networks, https://doi.org/10.1007/978-3-030-01731-6_1

deployment of different wireless technologies and the advanced vehicles with multiple network interfaces equipped, would allow in-vehicle users to access to different real-time services at anywhere anytime from any networks. Therefore, with a vehicular heterogeneous network formed by the cellular network and VANET, efficient network selection is crucial to ensuring vehicles' quality of service (QoS), avoiding network congestion and other performance degradation.

Researches for network selection in urban traffic environment are really few. In the context of future wireless networks, efficient network access system is required for vehicular users' real-time services provisioning [9–11]. We briefly introduce the hierarchical architecture of the heterogeneous vehicular networks as follows.

A hierarchical architecture is shown in Fig. 1.1. It consists of three layers: access network layer, data aggregation layer and application layer. In the access network layer, vehicles connect to cellular base station (eNB) through 3G/LTE or VANET base station (RSU) through DSRC technology. In the data aggregation layer, eNB and RSU are connected to the so-called central controller to access Internet. The aggregation center in the cloud aggregates data from vehicles and static sensors to estimate and predict the urban traffic. The cloud also connects with other service providers such that the traffic-related information can be fused out and provided in the application layer. Different services are then delivered to vehicles through the complementary resources of the cellular network and regional VANET.

In addition to the simple architecture, four components of the architecture are introduced to highlight the characteristics for real-time service provisioning.

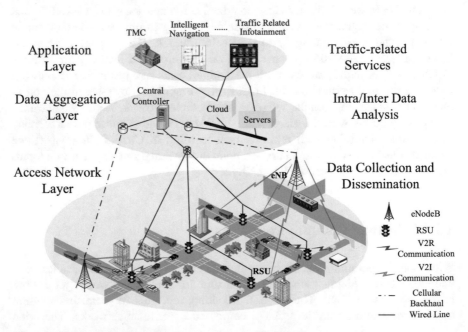

Fig. 1.1 Hierarchical architecture of vehicular networks for ITS

- **Infrastructure:** The infrastructures consist of eNBs and RSUs. The communication link between mobile devices and eNB is more stable than that between devices and RSU. Equipped with a wireless transceiver operating on DSRC, RSU has a smaller transmission range than eNB. But it provides higher rate and lower cost transmission for mobile devices.
- **Vehicle:** We do not discriminate what kind of mobile devices they are, but we care about which network they access to. Normally, smartphones can connect to cellular network through off-the-shelf 3G, while vehicles can connect to VANET and other vehicles through DSRC. We regard both smartphones and vehicles as vehicles.
- **Central controller:** Central controller connects with infrastructures and Internet. An access recommender console is applied in it to guide vehicles' network accessing based on the real-time traffic estimated through cloud calculation. It acts as an interface between physical network routers and network operators to specify network services.
- **Cloud:** As the computing center for traffic management center and other service providers, the cloud receives data from static traffic sensors and mobile sensors, and analyses them for traffic estimation and prediction. One key feature provided by cloud is the access guidance for vehicles in specific location according to traffic analysis and service requests.

With the development of autonomous driving, future vehicles will be equipped with various resources, including computing power, data storage and sensors. It is predicted that about 10^6 dhrystone million instructions executed per second (DMIPS) computing power is required by each vehicle to enable self-driving. These computing resources have huge potential to enhance the intelligence at the edge of the network, and can be integrated as Vehicular Edge Computing (VEC) systems for better utilization [12–14].

In VEC systems, vehicles and infrastructures like RSUs can contribute their computing resources to the network, while computation tasks are generated from various kinds of applications by the vehicular driving systems, on-board mobile devices and pedestrians. To be specific, in safety driving applications such as collective environmental perception and cooperative collision avoidance, the sensing data needs to be processed within tens of milliseconds [15]. By analyzing video recordings in real-time, vehicular crowd-sensing can help to monitor the road conditions and optimize traffic flows [16]. Computation tasks from other applications of mobile edge computing (MEC) [17], such as video stream analysis, augmented and virtual reality, IoT and tactile Internet, can also be offloaded to and processed by the VEC systems.

Compared with the MEC system, task offloading in the VEC faces more uncertainties due to the dynamic vehicular environment. First, the network topologies and the wireless channel states are time varying. Second, the density of vehicles is much higher than that of the static MEC servers, and the computing resources owned by vehicles are heterogeneous. How to allocate computing and communication resources of vehicles, in order to satisfy the delay requirements of tasks, is the key problem of task offloading in VEC systems.

In the VEC system, tasks can be either scheduled by a centralized controller, or offloaded in a distributed manner by each task generator. Some recent efforts are summarized as follows.

- **Centralized approaches:** A centralized VEC architecture is proposed in [14], inspired by the software-defined network. A centralized controller collects the state information of vehicles periodically, including location, velocity, moving direction, and computing resource occupation. Upon user requests, the controller predicts the instantaneous states of vehicles, and allocates the radio and computing resources to process the tasks. In [18], the task assignment problem is formulated based on the semi-Markov decision process. The objective is to minimize the average system cost, which is composed of the delay of tasks and the energy consumption saved at mobile devices. To further improve the reliability of computing services, task replication is introduced in [19], where task replicas are offloaded to multiple vehicles and processed at the same time.

  However, a major drawback of centralized task scheduling is that, the controller needs to govern the accurate state information of vehicles through frequent state updates, which brings high signaling overhead to the VEC systems. To reduce the signaling overhead, distributed approaches are investigated.

- **Distributed approaches:** In distributed approaches, task offloading decisions are made by each task generator individually. An autonomous offloading framework is proposed in [20], and then a task scheduling algorithm is designed based on ant colony optimization. Based on multi-armed bandit theory, a learning-based task offloading algorithm is proposed in [21], which enables task generators to learn the delay performance of other vehicles. The major challenge of distributed control is that vehicles may lack the global state information to make the optimal decision. The state information can be learnt by the task generators using online learning techniques, which will be discussed in detail in Chap. 3.

1.2 VANET Security

Due to time constraints, network scale, node mobility and volatility of the VANET, it's vulnerable to various attacks such as DoS, jamming, eavesdropping and spoofing attacks [22, 23], as shown in Fig. 1.2. For instance, the high mobility of OBUs and the large-scale network with infrastructures such as RSUs make the VANET vulnerable to jamming [24]. Smart jammers send faked or replayed signals with the goal to block the ongoing transmissions between the OBUs and the serving RSUs with flexible jamming strategies and strong radio sensing capabilities.

The global number of the connected vehicles increases rapidly, and vehicles are equipped with increasing amount of computing and storage resources. In order to improve the utilization of vehicle resources, the concept of vehicular cloud

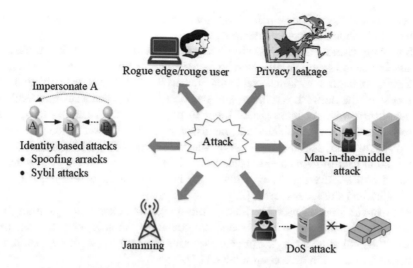

Fig. 1.2 Illustration of the threat model in VANET

computing (VCC) is proposed, in which vehicles can serve as vehicular cloud (VC) servers by sharing their surplus computing resources, and users such as other vehicles and pedestrians can offload computation tasks to them [21]. However, security is critical with increasing VCC usage and various security issues threaten the security of the VCC such as DoS, identity spoofing, modification repudiation, repudiation, Sybil attack, and information disclosure [25].

With the development of computation intensive applications such as augmented reality techniques, mobile devices such as smart watches and smartphones can offload the computation tasks to the OBUs in VANETs to utilize the computation, power, and storage resources of the edge node that can be an access point (AP) or even a laptop in the vehicle. The edge computing in VANETs has to address the rogue edge attacks. As a special type of spoofing attackers, an attacker can send spoofing signals to the mobile device claiming to be the serving edge node in the vehicle. Rogue edge attacks can result in the privacy leakage risks and further lead to man-in-the-middle attacks or denial-of-service (DoS) attacks [26]. We briefly review some important types of attacks as follows.

- **DoS** attackers overload the communication channels or break down the target server with superfluous requests to prevent the connected vehicles or vehicular users from obtaining services [22]. DDoS attackers use thousands of IP addresses to request VANET services, making it difficult for the server to distinguish the legitimate users from attackers [23].
- **Jamming:** A jammer sends faked signals to interrupt the ongoing radio transmissions of connected vehicles or vehicular users. Another goal of jammers is to deplete the bandwidth, energy, central processing unit (CPU) and memory

resources of the victim vehicles, inside-vehicular sensors and VC servers during their failed communication attempts [24].

- **Spoofing attacks/Rogue VC cloud/Rogue vehicular user/Sybil attacks:** An attacker sends spoofing signals to VC servers with cached chunks claiming the identity of another legitimate vehicular such as the MAC address to obtain illegal access of the network resources, and perform further attacks such as DoS and man-in-the-middle attacks [25, 27]. For example, an attacker claims to be a VC server to fool other vehicles and pedestrians in the area in rogue VC server attacks, or sends spoofing messages to the VC server with the identity of another vehicular user in rogue user attacks. In Sybil attacks as another type of identity-based attacks, a vehicular user claims to be multiple users and request more network and storage resources [25].
- **Man-in-the-middle attacks:** Man-in-the-middle attacker sends jamming and spoofing signals to fake VC servers, the connected vehicles or vehicular users with the goal of hijacking the private communication of the victim VC servers or vehicular users and even control them [27].
- **Eavesdropping:** Eavesdropper is located in a stopped or moving vehicle or in a false RSU and tends to illegally obtain access to confidential data [22].
- **Message falsification:** Attackers aim to provide erroneous information for the connected vehicles or vehicular users or modify the information sent by the VC server or vehicular users [22].
- **Information disclosure:** Malicious users acquire the specific information of a VANET system by revealing sensitive data, obtaining access to outsourced data and finding hidden content and functions [25].
- **Hardware tampering:** Vehicles can be perpetrated by other vehicles equipped with such as radar or global position system receivers and communication between the connected vehicles is disturbed [22].

1.3 Machine Learning Techniques in VANETs

With the development of machine learning (ML) and smart attacks, vehicular users and vehicles have to choose the defense policy and determine the key parameters in the security protocols for the tradeoff in the heterogenous and dynamic networks. It's challenging for vehicular users and vehicles to accurately estimate the current network and attack state in time due to the high mobility of OBUs and the large-scale network infrastructure with RSUs. For example, the rouge edge detection performance of the scheme in [26] is sensitive to the spoofing detection mode and the test threshold in the hypothesis test, which depends on both the radio propagation model and the spoofing model. Such information is unavailable for vehicles with high mobility, leading to a high false alarm rate or miss detection rate in the spoofing detection (Table 1.1).

Machine learning techniques including supervised learning, unsupervised learning, and reinforcement learning (RL) have been widely applied to improve network

Table 1.1 ML-based IoT security methods

Attacks	Machine learning techniques	Performance
DoS	Neural network [28]	Root-mean error
	Multivariate correlation analysis [29] Q-learning [30]	Detection accuracy
Jamming	PHC [24] Q-learning [31, 32] DQN [33]	SINR
	DQN [33]	Energy consumption
Spoofing	Q-learning [26, 34] Dyna-Q [34]	False alarm rate Miss detection rate
	SVM [35]	Classification accuracy
	DNN [36]	Average error rate Detection accuracy
Intrusion	Naive Bayes [37] K-NN [38]	Classification accuracy
	SVM [39] Neural network [39]	Classification accuracy False alarm rate Detection rate
Malware	Q/Dyna-Q/PDS [40]	Detection accuracy Detection latency
	Random forest [41] K-nearest neighbors [41]	Classification accuracy False positive rate True positive rate
Eavesdropping	Q-learning [42]	SINR
	Nonparametric Bayesian [43]	Proximity passing rate Secrecy data rate
Data congestion	K-means [44]	Communication delay Throughput Packet loss ratio
Misbehavior detection	Naive Bayes [45] Random forest classifier [45] Ensemble learning [45]	True positive rate False positive rate

security, such as authentication, access control, anti-jamming offloading and malware detections [24, 26, 28–30, 32, 33, 35–47].

- **Supervised learning** techniques such as support vector machine (SVM), naive Bayes, K-nearest neighbor (K-NN), neural network, deep neural network (DNN) and random forest can be used to label the network traffic or app traces of vehicles or vehicular users to build the classification or regression model [37]. For example, vehicles or vehicular users can use SVM to detect network intrusion [37] and spoofing attacks [35], apply K-NN in the network intrusion [38] and malware [41] detections, and utilize neural network to detect network intrusion [39] and DoS attacks [28]. Naive Bayes can be applied by vehicles or vehicular users in the intrusion detection [37], random forest classifier can be used to detect malwares [41] and K-means can be utilized to detect data congestion [44]. Vehicles or vehicular users with sufficient computation and memory resources can utilize DNN to detect spoofing attacks [36]. Naive Bayes, random forest classifier and ensemble learning are applied in the misbehavior detection [45].

- **Unsupervised learning** does not require labeled data in the supervised learning and investigates the similarity between the unlabeled data to cluster them into different groups [37]. For example, vehicles or vehicular users can use multivariate correlation analysis to detect DoS attacks [29] and apply the infinite Gaussian mixture model (IGMM) in the physical (PHY)-layer authentication with privacy protection [43].

- **Reinforcement learning** techniques such as Q-learning, Dyna-Q, post-decision state (PDS) [46] and deep Q-network (DQN) [47] enable a vehicle or vehicular user to choose the security protocols such as the spoofing detection mode as well as the key parameters against various attacks via trial-and-error [26]. For example, Q-learning as a model free RL technique has been used to improve the performance of the authentication against spoofing attacks [26], anti-jamming transmission [24, 32, 42], and malware detections [30, 40]. Vehicles or vehicular users can apply Dyna-Q in the authentication and malware detections [40], use PDS to detect malwares [40] and DQN in the anti-jamming transmission [33].

1.4 Summary

In Chap. 1, we have briefly reviewed the concepts in VANETs and investigated various attacks that throw serious threats to VANETs. As one of the most powerful tools, machine learning based methods have introduced to improve network security, and related work was reviewed.

References

1. W. Alasmary and W. Zhuang, "Mobility impact in ieee 802.11 p infrastructureless vehicular networks," *Ad Hoc Networks*, vol. 10, no. 2, pp. 222–230, Mar. 2012.
2. A. Abdrabou, B. Liang, and W. Zhuang, "Delay analysis for sparse vehicular sensor networks with reliability considerations," *IEEE Trans. Wireless Commun.*, vol. 12, no. 9, pp. 4402–4413, Sept. 2013.
3. H. A. Omar, W. Zhuang, and L. Li, "Vemac: A tdma-based mac protocol for reliable broadcast in vanets," *IEEE Trans. Mobile Computing*, vol. 12, no. 9, pp. 1724–1736, Sept. 2013.
4. S. Bharati and W. Zhuang, "Cah-mac: cooperative adhoc mac for vehicular networks," *IEEE J. Sel. Areas Commun.*, vol. 31, no. 9, pp. 470–479, Sept. 2013.
5. S. Bharati, W. Zhuang, L. V. Thanayankizil, and F. Bai, "Link-layer cooperation based on distributed tdma mac for vehicular networks," *IEEE Trans. Veh. Technol.*, vol. 66, no. 7, pp. 6415–6427, Jul. 2017.
6. K. Abboud, H. A. Omar, and W. Zhuang, "Interworking of dsrc and cellular network technologies for v2x communications: A survey," *IEEE Trans. Veh. Technol.*, vol. 65, no. 12, pp. 9457–9470, Dec. 2016.

7. S. Bharati and W. Zhuang, "Crb: Cooperative relay broadcasting for safety applications in vehicular networks," *IEEE Trans. Veh. Technol.*, vol. 65, no. 12, pp. 9542–9553, Dec. 2016.
8. K. Abboud and W. Zhuang, "Stochastic analysis of a single-hop communication link in vehicular ad hoc networks," *IEEE Trans. Intell. Transportation Syst.*, vol. 15, no. 5, pp. 2297–2307, Oct. 2014.
9. H. T. Cheng, H. Shan, and W. Zhuang, "Infotainment and road safety service support in vehicular networking: From a communication perspective," *Mechanical Systems and Signal Processing*, vol. 25, no. 6, pp. 2020–2038, Aug. 2011.
10. H. A. Omar, N. Lu, and W. Zhuang, "Wireless access technologies for vehicular network safety applications," *IEEE Network*, vol. 30, no. 4, pp. 22–26, Aug. 2016.
11. K. Abboud and W. Zhuang, "Impact of microscopic vehicle mobility on cluster-based routing overhead in vanets," *IEEE Trans. Veh. Technol.*, vol. 64, no. 12, pp. 5493–5502, Dec. 2015.
12. S. Abdelhamid, H. Hassanein, and G. Takahara, "Vehicle as a resource (VaaR)," *IEEE Netw.*, vol. 29, no. 1, pp. 12–17, Feb. 2015.
13. S. Bitam, A. Mellouk, and S. Zeadally, "VANET-cloud: A generic cloud computing model for vehicular ad hoc networks," *IEEE Wireless Commun.*, vol. 22, no. 1, pp. 96–102, Feb. 2015.
14. J. S. Choo, M. Kim, S. Pack, and G. Dan, "The software-defined vehicular cloud: A new level of sharing the road," *IEEE Veh. Technol. Mag.*, vol. 12, no. 2, pp. 78–88, Jun. 2017.
15. G. T. 22.886, "Study on enhancement of 3gpp support for 5g v2x services," V15.1.0, Mar. 2017.
16. J. Ni, A. Zhang, X. Lin, and X. S. Shen, "Security, privacy, and fairness in fog-based vehicular crowdsensing," *IEEE Commun. Mag.*, vol. 55, no. 6, pp. 146–152, Jun. 2017.
17. Y. Mao, C. You, J. Zhang, K. Huang, and K. B. Letaief, "A survey on mobile edge computing: The communication perspective," *IEEE Commun. Surveys & Tutorials*, vol. 19, no. 4, pp. 2322–2358, 2017.
18. K. Zheng, H. Meng, P. Chatzimisios, L. Lei, and X. Shen, "An SMDP-based resource allocation in vehicular cloud computing systems," *IEEE Trans. Ind. Electron.*, vol. 62, no. 12, pp. 7920–7928, Dec. 2015.
19. Z. Jiang, S. Zhou, X. Guo, and Z. Niu, "Task replication for deadline-constrained vehicular cloud computing: Optimal policy, performance analysis and implications on road traffic," *IEEE Internet Things J.*, vol. 5, no. 1, pp. 93–107, Feb. 2018.
20. J. Feng, Z. Liu, C. Wu, and Y. Ji, "AVE: autonomous vehicular edge computing framework with aco-based scheduling," *IEEE Trans. Veh. Technol.*, vol. 66, no. 12, pp. 10660–10675, Dec. 2017.
21. Y. Sun, X. Guo, S. Zhou, Z. Jiang, X. Liu, and Z. Niu, "Learning-based task offloading for vehicular cloud computing systems," in *Proc. IEEE Int'l Conf. Commun. (ICC)*, Kansas City, MO, May 2018.
22. R. G. Engoulou, M. Bellaïche, S. Pierre, and A. Quintero, "VANET security surveys," *Computer Commun.*, vol. 44, pp. 1–13, May 2014.
23. M. N. Mejri, J. Ben-Othman, and M. Hamdi, "Survey on VANET security challenges and possible cryptographic solutions," *Vehicular Commun.*, vol. 1, no. 2, pp. 53–66, Apr. 2014.
24. L. Xiao, X. Lu, D. Xu, Y. Tang, L. Wang, and W. Zhuang, "UAV relay in vanets against smart jamming with reinforcement learning," *IEEE Trans. Veh. Technol.*, vol. 67, no. 5, pp. 4087–4097, May 2018.
25. M. Whaiduzzaman, M. Sookhak, A. Gani, and R. Buyya, "A survey on vehicular cloud computing," *Journal of Network and Computer Applications*, vol. 40, pp. 325–344, Apr. 2014.
26. X. Lu, X. Wan, L. Xiao, Y. Tang, and W. Zhuang, "Learning-based rogue edge detection in vanets with ambient radio signals," in *IEEE Int'l Conf. Commun. (ICC)*, pp. 1–6, Kansas City, MO, May 2018.
27. L. Xiao and X. Wan and C. Dai and X. Du and X. Chen and M. Guizani, "Security in mobile edge caching with reinforcement learning," *IEEE Wireless Communications Magazine*, in press.

28. R. V. Kulkarni and G. K. Venayagamoorthy, "Neural network based secure media access control protocol for wireless sensor networks," in *Proc. Int'l Joint Conf. Neural Networks*, pp. 3437–3444, Atlanta, GA, Jun. 2009.

29. Z. Tan, A. Jamdagni, X. He, P. Nanda, and R. P. Liu, "A system for Denial-of-Service attack detection based on multivariate correlation analysis," *IEEE Trans. Parallel and Distributed Systems*, vol. 25, no. 2, pp. 447–456, May 2013.

30. Y. Li, D. E. Quevedo, S. Dey, and L. Shi, "SINR-based DoS attack on remote state estimation: A game-theoretic approach," *IEEE Trans. Control of Network Systems*, vol. 4, no. 3, pp. 632–642, Apr. 2016.

31. Y. Gwon, S. Dastangoo, C. Fossa, and H. Kung, "Competing mobile network game: Embracing anti-jamming and jamming strategies with reinforcement learning," in *Proc. IEEE Conf. Commun. and Network Security (CNS)*, pp. 28–36, National Harbor, MD, Oct. 2013.

32. M. A. Aref, S. K. Jayaweera, and S. Machuzak, "Multi-agent reinforcement learning based cognitive anti-jamming," in *Proc. IEEE Wireless Commun. and Networking Conf (WCNC)*, pp. 1–6, San Francisco, CA, Mar. 2017.

33. G. Han, L. Xiao, and H. V. Poor, "Two-dimensional anti-jamming communication based on deep reinforcement learning," in *IEEE Int'l Conf. Acoustics, Speech and Signal Processing*, pp. 2087–2091, New Orleans, LA, Mar. 2017.

34. L. Xiao, Y. Li, G. Han, G. Liu, and W. Zhuang, "PHY-layer spoofing detection with reinforcement learning in wireless networks," *IEEE Trans. Veh. Technol.*, vol. 65, no. 12, pp. 10037–10047, Dec. 2016.

35. M. Ozay, I. Esnaola, F. T. Yarman Vural, S. R. Kulkarni, and H. V. Poor, "Machine learning methods for attack detection in the smart grid," *IEEE Trans. Neural Networks and Learning Systems*, vol. 27, no. 8, pp. 1773–1786, Mar. 2015.

36. C. Shi, J. Liu, H. Liu, and Y. Chen, "Smart user authentication through actuation of daily activities leveraging WiFi-enabled IoT," in *Proc. ACM Int Symposium on Mobile Ad Hoc Networking and Computing (MobiHoc)*, pp. 1–10, Chennai, India, Jul. 2017.

37. M. Abu Alsheikh, S. Lin, D. Niyato, and H. P. Tan, "Machine learning in wireless sensor networks: Algorithms, strategies, and applications," *IEEE Commun. Surveys and Tutorials*, vol. 16, no. 4, pp. 1996–2018, Apr. 2014.

38. J. W. Branch, C. Giannella, B. Szymanski, R. Wolff, and H. Kargupta, "In-network outlier detection in wireless sensor networks," *Knowledge and Information Systems*, vol. 34, no. 1, pp. 23–54, Jan. 2013.

39. A. L. Buczak and E. Guven, "A survey of data mining and machine learning methods for cyber security intrusion detection," *IEEE Commun. Surveys and Tutorials*, vol. 18, no. 2, pp. 1153–1176, Oct. 2015.

40. L. Xiao, Y. Li, X. Huang, and X. J. Du, "Cloud-based malware detection game for mobile devices with offloading," *IEEE Trans. Mobile Computing*, vol. 16, no. 10, pp. 2742–2750, Oct. 2017.

41. F. A. Narudin, A. Feizollah, N. B. Anuar, and A. Gani, "Evaluation of machine learning classifiers for mobile malware detection," *Soft Computing*, vol. 20, no. 1, pp. 343–357, Jan. 2016.

42. L. Xiao, C. Xie, T. Chen, and H. Dai, "A mobile offloading game against smart attacks," *IEEE Access*, vol. 4, pp. 2281–2291, May 2016.

43. L. Xiao, Q. Yan, W. Lou, G. Chen, and Y. T. Hou, "Proximity-based security techniques for mobile users in wireless networks," *IEEE Trans. Information Forensics and Security*, vol. 8, no. 12, pp. 2089–2100, Oct. 2013.

44. N. Taherkhani and S. Pierre, "Centralized and localized data congestion control strategy for vehicular ad hoc networks using a machine learning clustering algorithm," *IEEE Trans. Intelligent Transportation Systems*, vol. 17, no. 11, pp. 3275–3285, Apr. 2016.

45. J. Grover, V. Laxmi, and M. S. Gaur, "Misbehavior detection based on ensemble learning in VANET," in *Int'l Conf. Advanced Computing, Networking and Security*, pp. 602–611, Surathkal, India, Dec. 2011.
46. X. He, H. Dai, and P. Ning, "Improving learning and adaptation in security games by exploiting information asymmetry," in *IEEE Conf. Computer Commun. (INFOCOM)*, pp. 1787–1795, Hongkong, China, May 2015.
47. V. Mnih and K. Kavukcuoglu and D. Silver and others, "Human-level control through deep reinforcement learning," *Nature*, vol. 518, no. 7540, pp. 529–533, Jan. 2015.

Chapter 2
Learning-Based Rogue Edge Detection in VANETs with Ambient Radio Signals

Rogue edge detection in VANETs is more challenging than the spoofing detection in indoor wireless networks due to the high mobility of onboard units and the large-scale network infrastructure with roadside units. In this chapter, we propose a physical-layer rogue edge detection scheme for VANETs according to the shared ambient radio signals observed during the same moving trace of the mobile device and the serving edge in the same vehicle. We also propose a privacy-preserving proximity-based security system for location-based services (LBS) in wireless networks, without requiring any pre-shared secret, trusted authority or public key infrastructure. In this scheme, the edge node under test has to send the physical properties of the ambient radio signals, including the received signal strength indicator (RSSI) of the ambient signals with the corresponding source media access control address during a given time slot. The mobile device can choose to compare the received ambient signal properties and its own record or apply the RSSI of the received signals to detect rogue edge attacks, and determines test threshold in the detection. Finally, we use a reinforcement learning technique to enable the mobile device to achieve the optimal detection policy in the dynamic VANET without being aware of the VANET model and the attack model.

2.1 Authentication in VANETs

In this section, the rogue edge attacks in VANETs are introduced. Meanwhile, to protect VANETs from suffering rogue edge attacks, physical-layer authentication techniques are proposed to detect rogue edges based on the physical properties of the ambient radio signals.

With the development of computation intensive applications such as augmented reality techniques, mobile devices such as smart watches and smartphones can offload the computation tasks to the onboard units in vehicular ad hoc networks to utilize the computation, power, and storage resources of the edge node that can

© Springer Nature Switzerland AG 2019

L. Xiao et al., *Learning-based VANET Communication and Security Techniques*, Wireless Networks, https://doi.org/10.1007/978-3-030-01731-6_2

be an access point or even a laptop in the vehicle. The edge computing significantly reduces the processing delay, increases the battery lifetime, and reduces the privacy risks at cloud. The edge computing in VANETs has to address the rogue edge attacks. As a special type of spoofing attackers, an attacker can send spoofing signals to the mobile device claiming to be the serving edge node in the vehicle. Rogue edge attacks can result in the privacy leakage risks and further lead to man-in-the-middle attacks or denial-of-service attacks [1].

Existing VANET authentication involves cryptography, trust, and certificate [2, 3]. However, mobile devices with restricted computation, storage and battery life in VANETs prefer to apply the lightweight authentication protocols that is robust against insider attacks. This issue can be addressed by physical-layer authentications, which depend on the spatial distribution of the PHY-layer properties of wireless transmissions, such as received signal strength (RSS) of the message signals, received signal strength indicator (RSSI) [4], the channel impulse responses of the radio channels [5–7], and the ambient radio signals [8, 9]. For instance, the Q-learning based authentication with the ambient radio signals (QAAS) scheme as proposed in [9] and the Q-learning based authentication with RSSI (QAR) scheme as developed in [4] enable a wireless device to detect spoofing in indoor wireless networks without being aware of the radio propagation model and the spoofing model. However, these PHY-layer authentication techniques usually have degraded detection performance of the edge offloading in VANETs that has higher mobility and more strict resource constraints than the indoor wireless networks [10].

The channel impulse response (CIR) based authentication presented in [11] uses the two-dimension CIR quantization scheme and the logarithmic likelihood ratio test to reduce the false alarm rate in wireless communications. The PHY-layer challenge-response authentication proposed in [12] for orthogonal frequency division multiplexing systems modulates the challenge and response messages to improve the authentication accuracy. The AP localization framework proposed in [13] exploits the channel state amplitude and phase to improve the localization accuracy for the RSS-based localization in wireless local area networks.

The ProxiMate system in [14] uses both ambient TV and FM radio signals for secure pairing and reduces the authentication error rates. The proximity test based on the infinite Gaussian mixture model in [8] exploits the PHY-layer features of the shared ambient radio signals of radio clients to improve the authentication accuracy in wireless networks. The active authentication proposed in [9] that utilizes the RSSIs of the ambient radio signals to detect spoofing mobile devices can provide accurate proximity test for indoor wireless networks.

The user-side authentication proposed in [15] that uses the global positioning system information to detect rogue APs in VANETs. The application-layer intrusion detection developed in [16] builds a hypothesis test to detect rogue nodes in VANETs. The certificateless aggregate signcryption in [3] uses RSUs for fog computing and storage units to reduce the computational cost of vehicular crowdsensing-based systems. The lightweight secure and privacy-preserving scheme in [17] improves the reliability of the vehicle-to-grid connection and reduces the total communication and computation costs in VANETs.

2.2   Rogue Edge Detection in VANETs

We consider a mobile device (Bob) and the serving edge node (Alice) in a vehicle. Bob aims to offload real-time data such as pictures, videos and virtual reality information to Alice, and has to determine whether the edge node under test is indeed Alice instead of a rogue node (Eve) outside the vehicle with the goal to spoof Bob.

Equipped with computation and caching resources, the edge node can reduce the processing delay and save the computation energy consumption of the mobile device. These radio nodes (Alice and Bob) can receive the signals sent by the radio sources nearby, such as the ambient AP, BS, RSU, and OBUs along their traces, as shown in Fig. 2.1. Both Bob and Alice can observe the RSSI and arrival time of the ambient radio signals and their MAC addresses, which can be used to detect rogue nodes. Once receiving the message that contains the computation results of Alice, Bob has to detect whether the message is indeed sent by Alice.

The ambient radio environment of Bob, Alice and Eve changes over time due to the high mobility of OBUs. Both Bob and Alice can extract and store the PHY-layer properties such as RSSIs, MAC address and arrival time of the ambient radio signals that are sent from multiple ambient radio sources such as the ambient OBUs (AOs), BSs, WiFi APs and RSUs at different locations. Bob asks the edge node under test to provide the PHY-layer feature trace information of the ambient radio signals before offloading data. Upon receiving Bob's request, the edge node under detection sends a message about their feature trace information to Bob, and Bob compares the features trace of the edge node with his trace to identify their shared M ambient packets in each time slot.

Without loss of generality, the vehicle that carries both the mobile device and the edge node is assumed to move along the road at a speed denoted by $v_1^{(k)}$ at time slot

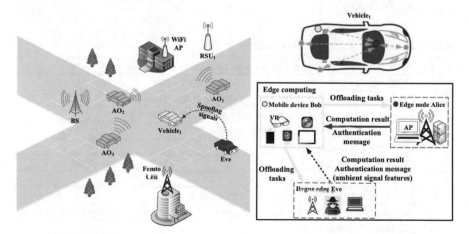

Fig. 2.1 Detection of rogue edge in VANETs based on the PHY-layer information of the ambient radio signals, in which mobile device Bob compares the PHY-layer information of the ambient radio signals received by the edge node under test with his copy to detect Eve

k. Let $r^{(k)}(i)$ denote the RSSI of the i-th packet sent by the edge node under test at time slot k, with $1 \leq i \leq N$, where N is the number of the packets of the edge node under test. Let $\hat{r}(i)$ denote the RSSI record of the i-th signal sent by Alice at time slot k.

The rogue edge (Eve) is assumed to be located in another vehicle with moving speed denoted by $v_2^{(k)}$ at time slot k. By sending a spoofing message with Alice's MAC address, Eve aims to fool Bob with faked edge computation results or illegal access advantages. In this PHY-layer authentication framework, Eve has to provide the PHY-layer feature of the ambient radio signals to Bob in the spoofing signals. Denote the probability that Eve sends a spoofing message at time slot k as $y^{(k)} \in [0, 1]$. Let $r_E^{(k)}(i)$ be the RSSI of the i-th packet sent by Eve at time slot k received by Bob.

Let $\hat{\mathbf{f}}^{(k)} = [\hat{r}_i^{(k)}, \hat{t}_i^{(k)}]_{1 \leq i \leq M}$ denote the feature trace record of Bob's ambient packets at time slot k, where $\hat{r}_i^{(k)}$ and $\hat{t}_i^{(k)}$ are the RSSI and arrival time of the i-th ambient radio signal monitored by Bob, where M is the number of the ambient signals assigned by Bob. Similarly, let $r_i^{(k)}$ and $t_i^{(k)}$ denote the RSSI, and arrival time of the i-th ambient packet monitored by the edge node under test. The ambient signal feature of the edge node under test denoted by $\mathbf{f}^{(k)} = [r_i^{(k)}, t_i^{(k)}]_{1 \leq i \leq M}$ at time slot k.

The path loss of a radio channel in the VANET for the link of distance d at time slot k is denoted by $L^{(k)}$. Similar to [18], we model the path loss in dB as

$$L^{(k)}[dB] = \bar{L}[dB] + 10n_0 \log(\frac{d}{d_0}) + \sigma \qquad (2.1)$$

where d_0 is the reference distance, $\bar{L}[dB]$ is the average large-scale path loss at d_0, n_0 is the path loss exponent and σ is the normal random variable corresponding to the shadowing fading. For ease of reference, some important notations are summarized in Table 2.1.

Table 2.1 Summary of symbols and notations

$r^{(k)}(i)$	RSSI of the i-th packet sent by the edge node under test at time slot k
$\hat{r}(i)$	RSSI record of the i-th signal sent by Alice at time slot k
$\hat{\mathbf{f}}^{(k)}$	Feature trace record of Bob's ambient packets at time slot k
$\mathbf{f}^{(k)}$	Ambient signal feature of the edge node under test at time slot k
$v_{1/2}^{(k)}$	Speed of the target/attack OBU at time slot k
$y^{(k)}$	Spoofing rate at time slot k
α	Learning rate of Bob
δ	Learning discount factor of Bob
$P_{f/m}^{(k)}$	False alarm rate/miss detection rate at time slot k
$\alpha_{f/m}$	Weight factors between the false alarm rate and the miss detection rate at time slot k
$u^{(k)}$	Utility of Bob at time slot k
$I^{(k)}$	Importance of Bob's sensing data at time slot k
$C_f^{(k)}$	Spoofing detection mode cost at time slot k

2.3   Proximity Test Based on Ambient Radio Signals

In this section, we propose a privacy-preserving proximity-based security system for location-based services (LBS) in wireless networks, without requiring any pre-shared secret, trusted authority or public key infrastructure. In this system, the proximity-based authentication and session key establishment are implemented based on spatial temporal location tags. Incorporating the unique physical features of the signals sent from multiple ambient radio sources, the location tags cannot be easily forged by attackers. More specifically, each radio client builds a public location tag according to the received signal strength indicators (RSSI), sequence numbers and MAC addresses of the ambient packets. Each client also keeps a secret location tag that consists of the packet arrival time information to generate the session keys. As clients never disclose their secret location tags, this system is robust against eavesdroppers and spoofers outside the proximity range. The system improves the authentication accuracy by introducing a nonparametric Bayesian method called infinite Gaussian mixture model in the proximity test and provides flexible proximity range control by taking into account multiple physical-layer features of various ambient radio sources. Moreover, the session key establishment strategy significantly increases the key generation rate by exploiting the packet arrival time of the ambient signals. The authentication accuracy and key generation rate are evaluated via experiments using laptops in typical indoor environments.

The pervasion of smartphones and social networks has boosted the rapid development of location-based services (LBS), such as the request of the nearest business and the location-based mobile advertising. Reliable and secure location-based services demand secure and accurate proximity tests, which allow radio users and/or service providers to determine whether a client is located within the same geographic region [19–22]. In order to support the business or financial oriented LBS services, proximity tests have to provide location privacy protection and location unforgeability [23–27].

Consequently, privacy-preserving proximity tests have recently drawn considerable research attention [28–34]. Based on the received signal strength (RSS) of a single radio source, many of the proximity tests suffer from the limited proximity range and the authentication accuracy is not high in both stationary and fast changing radio environments [31–33]. Moreover, a recent study has shown that the RSS-based strategies are vulnerable to man-in-the middle attacks [35]. To address this problem, Zheng et al. have proposed a location tag-based proximity test, which exploits the contents of ambient radio signals to improve the authentication accuracy and provides flexible range control [34]. However, the extraction of the packet contents in the proximity test not only engenders privacy leakage, but also increases the overall system overhead.

In typical indoor environments, a radio client can usually access multiple ambient radio sources, such as WiFi access points (APs), bluetooth devices and FM radios. Many off-the-shelf radio devices, such as laptops and smartphones, can readily extract the physical-layer features of the ambient signals, including the received signal strength indicator (RSSI) and the packet arrival time. Field tests have shown that clients in the same geographic area can observe a certain shared ambient signals, with approximately the same normalized packet arrival time and similar RSSIs. These physical-layer features do not directly disclose the client location and cannot be easily estimated and forged by a client outside the proximity [36]. Therefore, users can exploit the ambient radio signals to establish spatial temporal location tags and use the location tags to enhance security for LBS services.

In this section, we propose a proximity-based authentication and key generation strategy for radio clients, without involving any trusted authority, pre-shared secret or public key infrastructure. For simplicity, we assume that a radio client called Alice initiates the authentication and pairwise session key generation with clients in her proximity. A peer client called Bob responds to her request.[1] Both clients monitor their ambient radio signals at the frequency band during the time specified by Alice.

According to the physical-layer features of the signals sent by multiple ambient radio sources, Bob constructs and informs Alice his public location tag, which incorporates the RSSIs, sequence numbers (SN) and MAC addresses of the packets. Bob also builds and keeps a secret location tag, which consists of the packet arrival time sequence. Based on Bob's public location tag and her own measurements, Alice identifies their shared ambient packets and uses their features to derive the proximity evidence of Bob for both authentication and session key generation. Meanwhile, Alice informs Bob the indices of their shared packets in his secret location tag and helps him to generate his copy of the session key.

The authentication utilizes a nonparametric Bayesian method (NPB) called infinite Gaussian mixture model (IGMM)[37] to classify the RSSI data. This method avoids the "overfitting" problem and thus addresses the challenging issue of adjusting model complexity. The NPB method has shown its strength in the design of device fingerprints [38] and the detection of primary user emulation attacks in cognitive radio networks [39]. As an important alternative to deterministic inference such as expectation-maximization algorithm [40], the IGMM model is implemented in the proximity test to authenticate radio clients.

The proximity-based security system takes into account the packet loss due to the channel fading and interference, and can counteract various types of attacks. By hiding the packet arrival time sequence in the secret location tag, which is the basis of the session key and cannot be forged by malicious users, this scheme can efficiently address eavesdropping and spoofing attacks who are located outside the proximity range. Moreover, as public location tags do not disclose the client locations, location privacy is preserved for radio clients.

---

[1]This system can be directly extended to the case with Alice connecting to multiple peer clients.

Involving multiple ambient radio sources, the proximity test improves the authentication accuracy and obtains more flexible range control than those based on a single RSSI trace [31, 32]. Unlike the content-based location tag [34], the tag in this work consists of the physical-layer features of ambient signals, and thus avoids decoding the ambient signals. Therefore, this work is applicable to the case that the ambient packet decoding is not available or desirable, significantly reduces the computational overhead, and prevents privacy leakages.

As a location sharing method, proximity test enables the information sharing between users within a certain range. Related security issues have recently received significant attentions among researchers [30–33, 41–44]. In [30], a practical solution exploits the measured accelerometer data resulting from hand shaking to determine whether two smartphones are held by one hand.

For the proximity range exceeding a single hand, RSSI-based proximity tests were proposed in [31–33]. The proximity test in [31] calculates the Euclidean distance between the RSSIs of the shared ambient WiFi signals and applies a classifier called MultiBoost. The test in [32] relies on the feature of the peer client's signal. In [33], a secure pairing strategy exploits the amplitudes or phases of the shared ambient TV/FM radio signals to generate bits for the client pairs with longer proximity range. However, these methods are limited to the case where the distance between the radio clients is no more than a half wavelength away [33].

To achieve flexible range control, Zheng et al. proposed a private proximity test and secure cryto protocol, which applies the fuzzy extractors to extract secret keys and bloom filters to efficiently represent the location tags [34]. Inspired by this work, we propose a location tag-based security technique to further improve the performance, and some preliminary results were given in [45]. In this paper, we move forward to present the proximity-based security protocol that incorporates the proximity range control with fine granularity. We analyze the range control and security performance, and perform in-depth experiments to evaluate its performance such as the key generation rate and session key matching rate in typical indoor scenarios.

In this section, we consider two radio mobile clients called Alice and Bob, respectively, who are located in a certain geographic region. Without sharing any secret, trusted authority or public key infrastructure with Bob, Alice aims at initiating a proximity test and establishing a session key with him.

Both clients apply off-the-shelf radio devices, such as laptops and smartphones to extract the features of ambient radio signals, including the RSSIs, arrival time, source MAC addresses and sequence numbers (SN) of the packets. For simplicity, we take the 802.11 systems as an example in this section and consider the other types of radio sources in the later sections. In this system, each client monitors the ambient packets, which can be sent by access points (APs), over the frequency channel during the time specified by Alice, yielding a feature trace with N records.

As shown in [13, 16] and [28], a radio client in typical indoor environments can usually receive signals from *multiple* APs. For example, a stationary laptop in an experiment as later shown in Fig. 2.3 received signals from four APs in the 0.24 s time duration. Unlike [32], we utilize the ambient signals sent by multiple

APs instead of the testing packets sent by the clients or a single neighboring AP. In addition, clients have small chances to receive the same ambient packet sequence in the presence of multiple APs due to the path-loss and small-scale fading in radio propagation in typical indoor environments. Therefore, an attacker outside the proximity can rarely obtains all the shared ambient packets between Alice and Bob, and thus has difficulty predicting the exact arrival time sequence for their shared ambient packets.

In this work, Bob sends his temporal spatial location tag incorporating the trace information to Alice, and hence Alice obtains the RSSIs, MAC addresses and SN information of Bob's ambient signals. Let $rssi_i^A$, t_i^A, MAC_i^A and SN_i^A denote the RSSI, arrival time, MAC address and sequence number of the i-th ambient packet in Alice's feature trace, respectively, with $i = 1, \cdots, N$. Similarly, let $rssi_i^B$, t_i^B, MAC_i^B and SN_i^B represent the corresponding information monitored by Bob.

In this model, Alice initiates the proximity test, while Bob can be either an honest client in her proximity or an attacker outside the area. The proximity-based security techniques have to address the following types of adversary clients: (1) third-party eavesdroppers whose goal is to obtain the session key between Alice and Bob, (2) third-party attackers located outside the proximity range, who inject spoofed or replay signals in hopes of leading to a mismatched session key between Alice and Bob, and (3) Bob as an attacker who aims at illegally passing the proximity test although he is outside the proximity range specified by Alice. We will investigate the impacts of the other attackers in our future work. For ease of reference, the commonly used notations are summarized in Table 2.2.

Receivers in the proximity have similar RSSIs and approximately the same arrival time regarding their shared ambient radio signals. Without directly disclosing the clients' locations, these physical-layer features cannot be easily estimated and

Table 2.2 Summary of symbols and notations

$rssi_i^X$	RSSI of Packet i received by Client X
t_i^X	Arrival time of Packet i received by X
MAC_i^X	MAC addr. of Packet i received by X
SN_i^X	SN of Packet i received by X
$\mathbf{X}_i = [MAC_i^X, SN_i^X]$	MAC addr. & SN of Packet i received by X
N	Length of the trace recorded by Alice
D	Number of the ambient radio sources
$\mathbf{x} = [x_i]_{1 \leq i \leq n}$	Feature records obtained by Alice
$\mathbf{c} = [c_i]_{1 \leq i \leq n}$	Classification results of $\mathbf{x}$
Θ	Threshold to evaluate the Euclidean distance
ν	Proximity passing rate of Bob's records
Δ	Threshold to evaluate ν in the authentication
Υ	Rounding precision
Ξ	Threshold to evaluate the key generation rate in the authentication

thus be forged by clients outside the neighborhood [36]. Therefore, we propose a proximity-based authentication strategy for peer clients in wireless networks, where Alice decides whether Bob is in her proximity without violating his location privacy.

The proximity-based authentication is based on the similarity between the physical features of the shared ambient radio signals obtained by the radio clients. More specifically, Alice compares her trace with Bob's measurements extracted from his public location tag, according to a nonparametric Bayesian method (NPB) called infinite Gaussian mixture model (IGMM). Unlike the hypothesis tests such as maximum likelihood estimation, IGMM does not rely on the a priori knowledge of the input data model and works well even with uncertainty regarding the number of hidden classes and the data model [37]. In this authentication strategy, Alice classifies the RSSI information of the ambient signals from D APs to authenticate clients such as Bob.

2.3.1 IGMM-Based Proximity Test

According to Bob's location tag and her own feature trace, each with N records, Alice obtains a record vector $\mathbf{x}$ with $n = 2N$ feature records. For simplicity of denotation, we assume in this section that each record has only $D = 1$ dimension and $\mathbf{x} = [x_i]_{1 \leq i \leq n} \triangleq [rssi_1^A, \cdots, rssi_N^A, rssi_1^B, \cdots, rssi_N^B]$, where the first N elements correspond to Alice's trace. However, this method can be extended straightforwardly to the multivariate case with D features, where the gamma variables are replaced by Wishart random matrices and the normal variables become multinormal random vectors. As an example, the experiments that will be presented in Sect. 2.4.4 took into account the RSSI data of the signals sent by two ambient radio sources with $D = 2$.

The proximity test is based on the implementation of the IGMM model of $\mathbf{x}$ with the Markov chain Monte Carlo method called Gibbs sampling [40]. More specifically, first, we can use the finite Gaussian mixture model (FGMM) with k basis Gaussian distributions [37] to model the RSSI data x_i in Alice's record vector. In this model, the probability distribution function (pdf) of x_i is given by

$$p(x_i) = \sum_{l=1}^{k} \pi_l N(\mu_l, s_l^{-1}), \forall 1 \leq i \leq n, \tag{2.2}$$

where μ_l and s_l are the mean and precision of the l-th Gaussian distribution, respectively, and π_l is the mixing proportion [40] with $0 \leq \pi_l \leq 1$ and $\sum_{l=1}^{k} \pi_l = 1$.
The component means μ_l in Eq. (2.2) have the following Gaussian priors,

$$p(\mu_l | \lambda, r) \sim N(\lambda, r^{-1}), \tag{2.3}$$

where $\sim$ means "to be proportional to". Both the mean, λ, and precision, r, are hyperparameters with the same values for all the k components in FGMM. They have the following normal and gamma priors:

$$p(\lambda) \sim N(\mu_x, \sigma_x^2), \tag{2.4}$$

and

$$p(r) \sim G(1, \sigma_x^{-2}), \tag{2.5}$$

where μ_x and σ_x^2 are the mean and variance of the RSSI value x_i, respectively.

Let $\mathbf{c} = [c_i]_{1 \le i \le n}$ denote the classification labels of Alice's record vector $\mathbf{x}$, where c_i is the classification result of x_i, and $\mathbf{c}_{-i}$ incorporate the labels for the observations other than x_i. Following Bayesian principle, by (2.2) and (2.3), we can derive the posterior distribution of μ_l, conditioned on the classification results $\mathbf{c}$,

$$p(\mu_l|\mathbf{c}, \mathbf{x}, s_l, \lambda, r) \sim N(\frac{\bar{x}_l n_l s_l + \lambda r}{n_l s_l + r}, \frac{1}{n_l s_l + r}), \tag{2.6}$$

where $\bar{x}_l$ is the mean of the observations belonging to Class l that has n_l elements and is given by

$$\bar{x}_l = \frac{1}{n_l} \sum_{j:c_j=l} x_j. \tag{2.7}$$

Similar to the derivation in [37], according to (2.3), (2.4), (2.5), and (2.6), the posteriors of the hyperparameters, λ and r, are given by

$$p(\lambda|\mu_1, \cdots, \mu_k, r) \sim N(\frac{\mu_x \sigma_x^{-2} + r \sum_{l=1}^{k} \mu_l}{\sigma_x^{-2} + kr}, \frac{1}{\sigma_x^{-2} + kr}), \tag{2.8}$$

$$p(r|\mu_1, \cdots, \mu_k, \lambda) \sim G(k+1, \frac{k+1}{\sigma_x^2 + \sum_{l=1}^{k} (\mu_l - \lambda)^2}). \tag{2.9}$$

Similarly, the component precisions s_l in Eq. (2.2) have the Gamma priors as follows,

$$p(s_l|\beta, \omega) \sim G(\beta, \omega^{-1}), \tag{2.10}$$

whose shape β and mean ω^{-1} are hyperparameters in the FGMM model. Their priors have the following inverse Gamma and Gamma forms,

$$p(\beta^{-1}) \sim G(1, 1), \tag{2.11}$$

$$p(\omega) \sim G(1, \sigma_x^2). \tag{2.12}$$

By (2.2) and (2.10), we obtain the posterior of s_l as

$$p(s_l|\mathbf{c}, \mathbf{x}, \mu_l, \beta, \omega) \sim G(\beta + n_l, \frac{\beta + n_l}{\beta\omega + \sum_{j:c_l=l}(x_j - \mu_l)^2}). \tag{2.13}$$

Then, by combining Eqs. (2.10), (2.11), and (2.12) and after simplification, we have the following posteriors,

$$p(\omega|s_1, \cdots, s_k, \beta) \sim G(k\beta + 1, \frac{k\beta + 1}{\sigma_x^{-2} + \beta\sum_{j=l}^{k} s_j}), \tag{2.14}$$

$$p(\beta|s_1, \cdots, s_k, \omega) \sim \Gamma(\frac{\beta}{2})^{-k} e^{\frac{-1}{2\beta}} (\frac{\beta}{2})^{\frac{k\beta-3}{2}} \prod_{j=1}^{k} (\omega s_j)^{\frac{\beta}{2}} e^{-\frac{\beta s_j \omega}{2}}. \tag{2.15}$$

According to [37], the mixing proportion $\hat{\pi} = [\pi_l, \cdots, \pi_k]$ in Eq. (2.2) follows Dirichlet distribution, whose joint pdf is given by

$$p(\pi_1, \cdots, \pi_k|\alpha) = \frac{\Gamma(\alpha)\prod_{l=1}^{k}\pi_l^{\alpha/k-1}}{\Gamma(\alpha/k)^k}, \tag{2.16}$$

where $\Gamma(\cdot)$ is the Gamma function. The concentration parameter α in Eq. (2.16) has an inverse Gamma shape, and its prior and posterior can be written as

$$p(\alpha) \sim \alpha^{-3/2} \exp(-\frac{1}{2\alpha}), \tag{2.17}$$

$$p(\alpha|k, n) \sim \frac{\alpha^{k-3/2} \exp(-\frac{1}{2\alpha})\Gamma(\alpha)}{\Gamma(n+\alpha)}. \tag{2.18}$$

By using the standard Dirichlet integral and integrating out the mixing proportions, we have the prior of the indicators as the following,

$$p(c_1, \cdots, c_n|\alpha) = \int p(c_1, \cdots, c_n|\hat{\pi})p(\hat{\pi})d\pi_1 \cdots d\pi_k \tag{2.19}$$

$$= \frac{\Gamma(\alpha)}{\Gamma(n+\alpha)} \prod_{j=1}^{k} \frac{\Gamma(\alpha/k+n_j)}{\Gamma(\alpha/k)}, \tag{2.20}$$

where n_j is the number of data labelled with Class j.

Let $n_{-i,j}$ represent the number of data before x_l belonging to Class j, and $p(c_i = j|\mathbf{c}_{-i}, \alpha, n_{-i,j})$ denote the conditional prior probability for x_i in Class j. The infinite Gaussian mixture model can be viewed as an extreme case of FGMM with k in Eq. (2.20) approaching infinity. Consequently, if $n_{-i,j} > 0$, the conditional probability of c_i in the IGMM model can be simplified into

$$p(c_i = j | \mathbf{c}_{-i}, \alpha, n_{-i,j}) = \frac{n_{-i,j}}{n - 1 + \alpha}. \tag{2.21}$$

Otherwise, if no data has been assigned to Class j yet, i.e., $n_{-i,j} = 0$, the conditional probability of c_i in IGMM becomes

$$p(c_i = j | \mathbf{c}_{-i}, \alpha) = \frac{\alpha}{n - 1 + \alpha}. \tag{2.22}$$

According to Bayesian principle, we obtain the conditional posterior of the classification indicator as

$$p(c_i = j | \mathbf{c}_{-i}, \alpha, \mu_j, s_j) \sim p(c_i = j | \mathbf{c}_{-i}, \alpha) p(x_i | \mathbf{c}_{-i}, \mu_j, s_j). \tag{2.23}$$

The relationships among the hyperparameters (λ, r, β and ω), the input data $\mathbf{x}$ and the variables in the infinite Gaussian mixture model can be illustrated in the directed graph with plate notations in Fig. 2.2, where the rectangular block represents the repeated structure.

In the proximity test, we can apply the Gibbs sampling method to generate the random samples for the joint probability distributions given by the above formulas of the IGMM model. The classification indicators $\mathbf{c}$ can be calculated according to the observations $\mathbf{x}$. The number of distinct values in the resulting c_i indicates

Fig. 2.2 Directed graph with plate notations for the infinite Gaussian mixture model in the proximity test

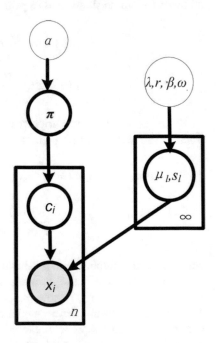

whether the recipient of the ambient signal is in the proximity of Alice. Ideally, all c_i take the same value if Bob is in the proximity with Alice, and take two different values if otherwise.

Detailed steps of the IGMM-based proximity test are illustrated in Algorithm 2.1, where NU is an integer that has to be set large enough to ensure accurate sampling

---

Algorithm 2.1 IGMM-based authentication

Require: RSSI measurements $\mathbf{x} = [x_i]_{1 \le i \le n}$
Ensure: Authentication result
  $k \leftarrow 1$
  $\mu_x \leftarrow E[\mathbf{x}]$, $\sigma_x^2 \leftarrow Var[\mathbf{x}]$
  $\lambda \leftarrow$ Eq. (2.4), $r \leftarrow$ (2.5), $\mu_l \leftarrow$ (2.3)
  $\beta \leftarrow$ (2.11), $\omega \leftarrow$ (2.12), $s_l \leftarrow$ (2.10)
  for $iter \leftarrow 0$ **to** NU **do**
    for $l \leftarrow 1$ **to** k **do**
      $\mu_l \leftarrow$ (2.6), $s_l \leftarrow$ (2.13)
    end for
    $\lambda \leftarrow$ (2.8), $r \leftarrow$ (2.9)
    $\beta \leftarrow$ (2.15), $\omega \leftarrow$ (2.14)
    $\alpha \leftarrow$ (2.18)
    for $i = 1$ **to** n **do**
      $c_i \leftarrow$ (2.23)
      if $c_i > k$ **then**
        Generate a new class c_i
        $\mu_{c_i} \leftarrow$ (2.3), $s_{c_i} \leftarrow$ (2.10)
      end if
      Update $\mathbf{c}$ by deleting empty classes
      $k \leftarrow$ Number of distinct components in $\mathbf{c}$
    end for
  end for
  Update $\mathbf{c}$ by combining the classes whose centroid Euclidean distance is less than Θ
  $C_A \leftarrow$ (2.24)
  $j \leftarrow 0$
  for $i \leftarrow N + 1$ **to** $2N$ **do**
    if $c_i = C_A$ **then**
      Alice accepts the packet, $j + +$
    end if
  end for
  Proximity passing rate $v \leftarrow j/N$
  if $v > \Delta$ **then**
    Bob passes the authentication
  else
    Bob fails the authentication
  end if

---

for the IGMM model. In addition, the system parameter N has to be less than the maximum value of sequence number of the specified ambient signals to avoid packet aliasing.

2.3.2 Post-IGMM Process

As radio signals in typical radio environments usually have time-variant RSSIs, a post-IGMM process is proposed to address slight channel time variations. This process combines the classes resulting from the IGMM-based proximity test, if they are close to each other. More specifically, if the Euclidean distance of the centroids of two classes is below a threshold denoted as Θ, these two classes are joined together. We now take Class i and j for instance. If $\| E_{c_l=i}[x_l] - E_{c_l=j}[x_l] \| < \Theta$, we combine these data and update the labels $c_l \in \{i, j\}$ with $\min(i, j)$, $\forall 1 \leq l \leq n$. Then the empty class is deleted by reducing c_l by one if their original value $c_l > \max(i, j)$.

Next, we apply the majority rule to process Alice's trace with N records and calculate their new label C_A by the following,

$$C_A = \arg\max_{c \in \mathbf{c}} \sum_{i=1}^{N} \delta(c_i - c), \qquad (2.24)$$

where $\delta(\cdot)$ is the discrete delta function. Alice accepts the data whose label equals C_A. We define the proximity passing rate denoted with ν as the ratio of Bob's records that pass the proximity test after the majority rule. Bob passes the proximity test, if the passing rate of his monitored ambient packets exceeds a threshold Δ, i.e., $\nu > \Delta$.

The above IGMM-based authentication strategy is summarized in Algorithm 2.1. Besides this RSSI-based strategy, we also provide another authentication strategy that exploits the packet arrival time to achieve a larger proximity range. More specifically, as the key generation rate of the strategy given by Algorithm 2.2 contains proximity information, we can utilize this information for authentication purpose. More details will be provided in Sect. 2.3.3.

2.3.3 Session Key Establishment

Note that clients receive the shared ambient radio packets approximately at the same time. Hence they can exploit the arrival time of the packets to establish pair-wise session keys without requiring any pre-shared secret, trusted authority or public key infrastructure. To this end, Alice initiates the process by broadcasting her key establishment policy. Upon receiving the policy, radio clients in the proximity

---

Algorithm 2.2 Session key generation

Require:

$\underline{A} = [\mathbf{A}_i]_{1 \le i \le N}^T, \mathbf{A}_i = [MAC_i^A, SN_i^A]$

$\underline{B} = [\mathbf{B}_i]_{1 \le i \le N}^T, \mathbf{B}_i = [MAC_i^B, SN_i^B]$

t_i^A and t_i^B: packet arrival time, $1 \le i \le N$

Υ: Rounding precision

Ensure: Session Key, $\mathbf{K}_A$ and $\mathbf{K}_B$

$\quad \mathbf{I} \leftarrow \{i | \exists j, 0 \le i, j \le N, \mathbf{A}_i = \mathbf{B}_j\}$

$\quad \mathbf{J} \leftarrow \{j | \exists i, 0 \le i, j \le N, \mathbf{A}_i = \mathbf{B}_j\}$

$\quad$ Alice sends $\mathbf{J}$ to Bob

$\quad t_a \leftarrow t_1^A, t_b \leftarrow t_1^B$

$\quad$ **for** $i \leftarrow 1$ **to** N **do**

$\quad\quad T_i^A \leftarrow round(t_i^A - t_a, 10^{-\Upsilon})$

$\quad\quad T_i^B \leftarrow round(t_i^B - t_b, 10^{-\Upsilon})$

$\quad$ **end for**

$\quad \mathbf{K}_A \leftarrow [T_i^A]_{i \in \mathbf{I}}$

$\quad \mathbf{K}_B \leftarrow [T_i^B]_{i \in \mathbf{J}}$

---

including Bob monitor the ambient signals accordingly and build their spatial temporal location tags by extracting the physical-layer features of the signals.

Each location tag consists of two parts: a secret location tag that incorporates the packet arrival time information and is kept by the client, and the public location tag that informs Alice the RSSIs for authentication and the MAC addresses and SNs to identify ambient packets.[2] To counteract the difference between the secret location tag between clients due to the transmission over air, the measured packet arrival time is rounded according to a properly chosen rounding precision. The rounding precision denoted with Υ is a tradeoff between the key generation speed and the key matching rate between clients.

For simplicity of notation, we take the key establishment between Alice and Bob as an example. Define $\underline{A} \triangleq [\mathbf{A}_i]_{1 \le i \le N}$ and $\underline{B} \triangleq [\mathbf{B}_i]_{1 \le i \le N}$, where $\mathbf{A}_i \triangleq [MAC_i^A, SN_i^A]$ and $\mathbf{B}_i \triangleq [MAC_i^B, SN_i^B]$. Bob's secret location tag contains t_i^B, $1 \le i \le N$, and his public location tag consists of $\underline{B}$, i.e., the MAC addresses and SNs of his ambient signals.

To address the transmission time, both Alice and Bob round the packet arrival time according to Υ. In particular, we have $T_i^A \triangleq round(t_i^A - t_a, 10^{-\Upsilon})$ and $T_i^B \triangleq round(t_i^B - t_b, 10^{-\Upsilon})$, where $t_a = t_1^A$ and $t_b = t_1^B$ are introduced to address the clock difference between these radio devices. The selections of $\Upsilon = 1, 2$ and 3 correspond to the rounding of the time information to the order of 0.1 s, 0.01 s and 1 ms, respectively. Experimental results show that $\Upsilon = 2$ is a reasonable choice for ambient WiFi signals.

The session key generation process is presented in Algorithm 2.2. Upon receiving Bob's public location tag, Alice compares it with her trace to identify their shared

---

[2]The duration is assumed to be short enough to avoid the reuse of SN for a given radio source.

ambient packets. As a result, Alice obtains their indices in her trace and Bob's trace, given by $\mathbf{I} = \{i | \exists j, 0 \leq i, j \leq N, \mathbf{A}_i = \mathbf{B}_j\}$, and $\mathbf{J} = \{j | \exists i, 0 \leq i, j \leq N, \mathbf{A}_i = \mathbf{B}_j\}$, respectively. Then Alice sends $\mathbf{J}$ to Bob.

In the next step, Alice generates her session key $\mathbf{K}_A$ based on the arrival time of their shared packets, i.e., $\mathbf{K}_A = [T_i^A]_{i \in \mathbf{I}}$. Similarly, Bob uses $\mathbf{J}$ to find their shared packets in his secret location tag and derives his session key with $\mathbf{K}_B = [T_i^B]_{i \in \mathbf{J}}$. The proposed key establishment process is summarized in Algorithm 2.2. We can see that this strategy has low complexity and is easy to implement.

2.4   Authentication Based on Ambient Radio Signals

In this section, we present the proximity-based security protocol and discuss related issues such as the proximity range control and the performance against various types of attackers.

2.4.1   Proximity Range Control

In this system, Alice can control the proximity range by choosing appropriate ambient radio sources and signal features at multiple levels. First, as shown in Table 2.3, radio devices such as smartphones and laptops can access multiple radio sources with various coverage ranges and frequency bands. By switching her frequency bands, Alice chooses the radio sources whose coverage ranges are larger than the proximity range. For example, Alice can use FM radio signals for the proximity range of several miles, and choose WiFi or bluetooth signals if contacting with clients within the same room.

Second, the range control can also be achieved by selecting suitable physical-layer features, since the features have different coherent spacial distances. For example, Alice and Bob usually obtain different RSSIs if their distance is greater than a half wavelength, which is around several centimeters for WiFi sources. On the other hand, two clients can receive a shared packet approximately at the same time, even if they are more than 30 m away. Therefore, we perform a fine-range proximity test by taking into account the RSSIs of the ambient signals and implement a large-range test based on the normalized packet arrival time.

Table 2.3 Range control by selecting different ambient radio sources in the proximity-based security system

System	Bluetooth	WLAN	GSM	FM radio
Frequency (Hz)	2.4G	2.4,5G	0.9/1.8G	87.5–108M
Range (m)	~10	~35	~30k	>100k

The RSSI-based proximity test has been given in Algorithm 2.1, where the range granularity is determined by the thresholds in the post-IGMM process. In general, the range granularity decreases with the threshold Θ. The thresholds are determined according to the proximity range via training in the similar environments.

Moreover, we also propose an authentication strategy by exploiting the packet arrival time feature of the ambient signals. As shown in Fig. 2.8b, the key generation rate of Algorithm 2.2 decreases smoothly and approximately monotonically with the client distance. Therefore, Alice can evaluate the key generation performance of Algorithm 2.2 to perform the proximity-based authentication. More specifically, Alice compares her key generation rate with a threshold denoted as Ξ: she believes that Bob is in her proximity if her key generation rate is higher than Ξ, and rejects Bob if otherwise.

As will be shown in the experimental results in Sect. 2.4.4, the packet arrival time-based authentication strategy can control the proximity range more flexibly. In that strategy, the coverage range that is more than 50 m for WiFi signals is much larger than the proximity range of the method in [33], which is around several centimeters. On the other hand, if Alice's proximity range is short, Algorithm 2.1 that is based on RSSIs achieves a higher authentication accuracy.

2.4.2   Proximity-Based Security Protocol

By integrating the authentication and key generation process, we build a proximity-based security protocol for mobile users in wireless networks. As illustrated in Fig. 2.3, this protocol consists of the following steps:

1. According to the desired proximity range, Alice decides and broadcasts her proximity test policy, including the frequency channel, the time duration and the features to monitor the ambient signals.
2. Upon receiving Alice's request, Bob measures the features of the packets as Alice specified. Both clients extract and store the RSSIs, arrival time, MAC addresses and sequence numbers of their ambient packets, i.e., $rssi_i^X, t_i^X, MAC_i^X$ and SN_i^X, with $1 \leq i \leq N$.
3. Bob builds a location tag, sends Alice his public location tag, and keeps his secret location tag.
4. Alice authenticates Bob according to Algorithm 2.1.
5. Alice compares Bob's public location tag with her trace to identify their shared packets. Following Algorithm 2.2, Alice builds a session key, $\mathbf{K}_A$, and informs Bob the indices of their shared packets in his trace, $\mathbf{J}$.
6. Based on his secret location tag and the indices $\mathbf{J}$, Bob generates his session key, $\mathbf{K}_B$, following Algorithm 2.2.

In the above handshake process, error correction coding such as BCH can be applied to counteract the transmission errors due to channel fading and interference. In addition, because of the different ambient radio environments and packet loss

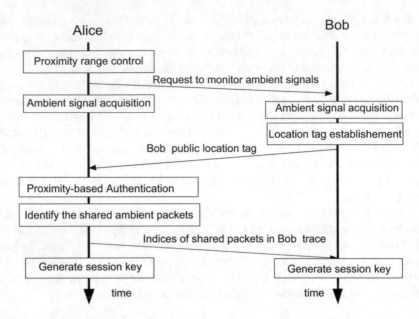

Fig. 2.3 Flowchart of the proximity-based security system based on ambient radio signals

---

Algorithm 2.3 Simplified proximity-based authentication

Require: RSSI measurements $\mathbf{x} = [x_i]_{1 \le i \le n}$
Ensure: Authentication result
$\quad c_i = i, \forall 1 \le i \le n$
$\quad$ Update $\mathbf{c}$ by combining the classes whose centroid Euclidean distance is less than Θ
$\quad C_A \leftarrow$ (2.24)
$\quad j \leftarrow 0$
$\quad$ **for** $i \leftarrow N + 1$ **to** $2N$ **do**
$\quad\quad$ **if** $c_i = C_A$ **then**
$\quad\quad\quad$ Alice accepts the packet, $j + +$
$\quad\quad$ **end if**
$\quad$ **end for**
$\quad$ Proximity passing rate $\nu \leftarrow j/N$
$\quad$ **if** $\nu > \Delta$ **then**
$\quad\quad$ Bob passes the authentication
$\quad$ **else**
$\quad\quad$ Bob fails the authentication
$\quad$ **end if**

---

rates, clients usually take different time to obtain a given number of ambient packets. Due to this problem, the proposed key generation strategy solely relies on the same shared packets between Alice and Bob and thus provides a certain degree of robustness against packet loss.

As comparison, we propose a simplified version of the proximity-based authentication strategy. As described in Algorithm 2.3, this strategy is based on the RSSI information of the ambient radio signals and applies the Euclidean distance method for classification. By skipping the IGMM process of Algorithm 2.1, this strategy reduces the system overhead and complexity.

2.4.3 Security and Performance Analysis

The proximity-based security technique is robust against the eavesdropper whose goal is to locate clients. As shown in Fig. 2.3, all that eavesdroppers can capture are the indices **J** and Bob's public location tag that consists of the RSSIs, SNs and MAC addresses of the ambient packets. Since neither of them directly discloses Bob's location, this system can protect the location privacy.

As shown in [35], existing key generation strategies that are based on the RSSI and channel impulse response (CIR) [46–50] or the phase [51] are vulnerable to the man-in-the-middle attacks. For instance, eavesdroppers can reveal 40% to 50% of the keys, and attackers can sabotage the key agreements with 95% confidence by injecting spoofing signals during less than 4% of the overall communication duration [35].

Fortunately, man-in-the-middle attacks out of the proximity can be addressed in the proposed key establishment system by exploiting the packet arrival time. Because of the packet loss due to the channel fading that decorrelates fast over space, it is highly challenging for an attacker outside the proximity to estimate the exact ambient packet arrival time sequence of a client, if there are *multiple ambient radio sources*, which is true in most indoor environments. For example, Fig. 2.4 presents a packet arrival sequence captured by a client with a wireless adapter in an experiment, showing the difficulty in estimating the exact SN sequence over time and thus the corresponding packet arrival time. This system never broadcasts the packet arrival time information over the air. Therefore, eavesdroppers outside the proximity cannot derive the pairwise session key between Alice and Bob.

Next, we consider attackers who spoof ambient radio sources by injecting faked or replay signals in hopes of significantly increasing the key disagreement rate between Alice and Bob in Algorithm 2.2. Note that the actual ambient radio source and the attacker usually result in different RSSIs in their signals due to distinct locations. Therefore, the faked packets can hardly pass the proposed proximity-based authentication, and thus are discarded in the session key generation using Algorithm 2.2. In addition, even with the knowledge of the past RSSI information, attackers still have difficulty in estimating the current RSSI obtained by the radio client due to the random time variation of RSSIs. Consequently, the proposed authentication strategy can also filter out the relayed messages.

Finally, compared with the time-variant RSSI or CIR, the packet arrival time has much higher entropy and is less sensitive to the radio propagation pattern. Therefore, as will be shown in the experimental results, this system can generate session keys much faster, and control the proximity range more flexible than the RSS-based key generation strategies such as [50]. Moreover, by introducing the IGMM method and exploiting the packet arrival time information, this security system provides more

No.	Time	Source	RSSI	Info
1	0.000000	Cisco_6c:66:50	10 dB	Beacon frame, SN=46, FN=0, Flags=........C, BI=100, SSID=UVA_WLAN, Name="NO\
2	0.005905	Cisco_6c:57:f0	24 dB	Beacon frame, SN=2374, FN=0, Flags=........C, BI=100, SSID=UVA_WLAN, Name="\
3	0.012906	Cisco_6c:67:41	33 dB	Beacon frame, SN=2942, FN=0, Flags=........C, BI=100, SSID=VT-Wireless, Name
4	0.014344	Cisco_6a:7f:00	26 dB	Beacon frame, SN=1471, FN=0, Flags=........C, BI=100, SSID=UVA_WLAN, Name="\
5	0.021717	Cisco_6c:69:00	48 dB	Beacon frame, SN=22, FN=0, Flags=........C, BI=100, SSID=UVA_WLAN, Name="NO\
6	0.034938	Cisco_6c:66:52	11 dB	Beacon frame, SN=47, FN=0, Flags=........C, BI=100, SSID=VT_WLAN, Name="NOV-
7	0.036432	Cisco_6c:65:e0	26 dB	Beacon frame, SN=3705, FN=0, Flags=........C, BI=100, SSID=UVA_WLAN, Name="\
8	0.040832	Cisco_6c:57:f2	24 dB	Beacon frame, SN=2375, FN=0, Flags=........C, BI=100, SSID=VT_WLAN, Name="NC
9	0.046431	Cisco_6c:67:40	34 dB	Beacon frame, SN=2943, FN=0, Flags=........C, BI=100, SSID=UVA_WLAN, Name="\
10	0.047777	Cisco_6a:7f:02	26 dB	Beacon frame, SN=1472, FN=0, Flags=........C, BI=100, SSID=VT_WLAN, Name="NC
11	0.069179	Cisco_6c:66:51	8 dB	Beacon frame, SN=48, FN=0, Flags=........C, BI=100, SSID=VT-Wireless, Name='
12	0.071052	Cisco_6c:65:e2	27 dB	Beacon frame, SN=3706, FN=0, Flags=........C, BI=100, SSID=VT_WLAN, Name="NC
13	0.076401	Cisco_6c:57:f1	23 dB	Beacon frame, SN=2376, FN=0, Flags=........C, BI=100, SSID=VT-Wireless, Name
14	0.081208	Cisco_6a:7f:01	28 dB	Beacon frame, SN=1473, FN=0, Flags=........C, BI=100, SSID=VT-Wireless, Name
15	0.082926	Cisco_6c:67:42	34 dB	Beacon frame, SN=2944, FN=0, Flags=........C, BI=100, SSID=VT_WLAN, Name="NC
16	0.102443	Cisco_6c:66:50	9 dB	Beacon frame, SN=49, FN=0, Flags=........C, BI=100, SSID=UVA_WLAN, Name="NO\
17	0.108227	Cisco_6c:57:f0	22 dB	Beacon frame, SN=2377, FN=0, Flags=........C, BI=100, SSID=UVA_WLAN, Name="\
18	0.115270	Cisco_6c:67:41	35 dB	Beacon frame, SN=2945, FN=0, Flags=........C, BI=100, SSID=VT-Wireless, Name
19	0.116731	Cisco_6a:7f:00	28 dB	Beacon frame, SN=1474, FN=0, Flags=........C, BI=100, SSID=UVA_WLAN, Name="\
20	0.137316	Cisco_6c:66:52	10 dB	Beacon frame, SN=50, FN=0, Flags=........C, BI=100, SSID=VT_WLAN, Name="NOV-
21	0.138766	Cisco_6c:65:e0	27 dB	Beacon frame, SN=3708, FN=0, Flags=........C, BI=100, SSID=VT_WLAN, Name="\
22	0.140195	Cisco_6c:5b:62	41 dB	Beacon frame, SN=2622, FN=0, Flags=........C, BI=100, SSID=VT_WLAN, Name="NC
23	0.143085	Cisco_6c:57:f2	23 dB	Beacon frame, SN=2378, FN=0, Flags=........C, BI=100, SSID=VT_WLAN, Name="NC
24	0.148639	Cisco_6c:67:40	36 dB	Beacon frame, SN=2946, FN=0, Flags=........C, BI=100, SSID=UVA_WLAN, Name="\
25	0.171385	Cisco_6c:66:51	10 dB	Beacon frame, SN=51, FN=0, Flags=........C, BI=100, SSID=VT-Wireless, Name='
26	0.183655	Cisco_6a:7f:01	27 dB	Beacon frame, SN=1476, FN=0, Flags=........C, BI=100, SSID=VT-Wireless, Name
27	0.185282	Cisco_6c:67:42	37 dB	Beacon frame, SN=2947, FN=0, Flags=........C, BI=100, SSID=VT_WLAN, Name="NC
28	0.204759	Cisco_6c:66:50	7 dB	Beacon frame, SN=52, FN=0, Flags=........C, BI=100, SSID=UVA_WLAN, Name="NO\
29	0.210682	Cisco_6c:57:f0	23 dB	Beacon frame, SN=2380, FN=0, Flags=........C, BI=100, SSID=UVA_WLAN, Name="\
30	0.217784	Cisco_6c:67:41	36 dB	Beacon frame, SN=2948, FN=0, Flags=........C, BI=100, SSID=VT-Wireless, Name
31	0.219406	Cisco_6a:7f:00	27 dB	Beacon frame, SN=1477, FN=0, Flags=........C, BI=100, SSID=UVA_WLAN, Name="\
32	0.239625	Cisco_6c:66:52	8 dB	Beacon frame, SN=53, FN=0, Flags=........C, BI=100, SSID=VT_WLAN, Name="NOV-
33	0.241278	Cisco_6c:65:e0	25 dB	Beacon frame, SN=3711, FN=0, Flags=........C, BI=100, SSID=UVA_WLAN, Name="\
34	0.245495	Cisco_6c:57:f2	23 dB	Beacon frame, SN=2381, FN=0, Flags=........C, BI=100, SSID=VT_WLAN, Name="NC

Fig. 2.4 Sequence numbers and MAC addresses of the ambient WiFi signals captured by wireless adapters *AirPcap Nx* and open-source packet analyzers *Wireshark* in an experiment

accurate authentication with flexible range control for larger coverage area than the strategies in [31–33]. More in-depth analysis of the security performance will be performed in our future work.

2.4.4 Experimental Results

We performed experiments in Virginia Tech Northern Virginia Center to evaluate the performance of this system. As shown in Figs. 2.5 and 2.9, two laptops acting as Alice and Bob, respectively, were placed in different locations in the 2nd floor of the building. Utilizing wireless adapters *AirPcap Nx* and open-source packet analyzers *Wireshark*, both laptops simultaneously captured the ambient WiFi signals.

The settings of the first experiment with 17 scenarios are shown in Fig. 2.5, where Bob was placed in different locations along the hallway. Both clients recorded the RSSI from 2 ambient WiFi APs. An example of the difference between the ambient RSSI vectors obtained by Alice and Bob is presented in Fig. 2.5b, showing that the average RSSI difference often increases with the distance between Alice and Bob, especially when the distance between Alice and Bob is less than 15 m. On the other hand, their relationship is in general complicated, as RSSI also depends on the transmitter location and the specific radio environment.

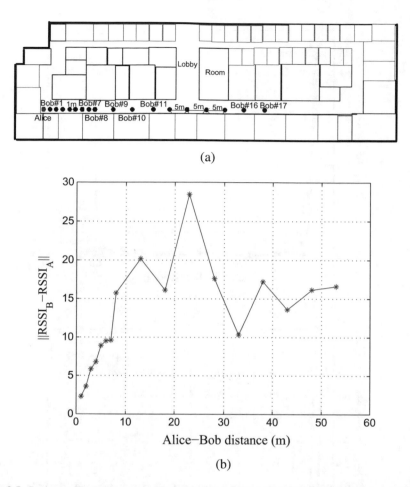

(a)

(b)

Fig. 2.5 Settings of Experiment 1 performed in Virginia Tech Northern Virginia Center. (**a**) Client placements. (**b**) Example of the average difference between the RSSI vectors observed by Alice and Bob

We calculated two metrics to evaluate the authentication performance: (1) Type 1 error rate, also known as false alarm rate or false rejection rate, is the probability that Alice rejects the packet from a client in her proximity by mistake; and (2) Type 2 error rate, or the false acceptance rate, is the probability to falsely accept a packet sent by a client outside her proximity.

We present the probability for Bob to pass the proximity test by Alice in different scenarios for both Algorithm 2.1 with the threshold $\Theta = 7.5$ and Algorithm 2.3. As illustrated in Fig. 2.6a, Alice can accurately determine whether Bob is in her proximity with the 4 m proximity range. For example, the false rejection rate of Algorithm 2.1 is very small if the Alice-Bob distance is less than 3 m. In this case, the false acceptance rate is less than 5% when the distance between Alice and Bob

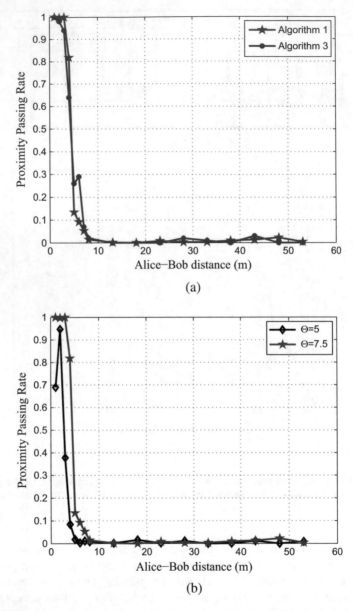

Fig. 2.6 Performance of the proximity-based authentication in Experiment 1. (**a**) Proximity passing rate of Algorithms 2.1 and 2.3 with $\Theta = 7.5$. (**b**) Proximity passing rate of Algorithm 2.1 with $\Theta = 5$ and 7.5

is larger than 6 m, and is very small when the Alice-Bob distance is more than 10 m. We also provide the performance of Algorithm 2.1 with different Θ in Fig. 2.6b, showing that $\Theta = 7.5$ is a good heuristic choice for the authentication with the 4 m proximity range.

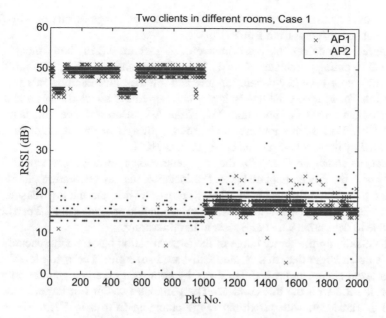

Fig. 2.7 RSSI trace with $D = 2$ and $N = 2$, as the input of Algorithm 2.1

Compared with Algorithm 2.3, the NPB-based strategy, Algorithm 2.1, is more stable in both the rejection region and the passing region, and has a narrower transition region. For example, the Type 1 error rate of Algorithm 2.1 is more than 5% lower than Algorithm 2.3, when the Alice-Bob distance is 2 m and the proximity range is 3 m. Meanwhile, Algorithm 2.1 rejects the clients outside the proximity more accurately. For instance, the Type 2 error of Algorithm 2.1 is about 20% lower than Algorithm 2.3, when the Alice-Bob distance is 6 m and the proximity range is 3 m. On the other hand, Algorithm 2.3 also works well when the Alice-Bob distance is much larger than the proximity range (e.g., the proximity range and Alice-Bob distance are 3 and 40 m, respectively), as shown in Fig. 2.6a.

In each scenario, clients extracted the RSSI, packet arrival time, SN and MAC addresses of the ambient beacon frames at 2.417 GHz, and recorded the trace for one minute. Both clients recorded the RSSIs from $D = 2$ ambient WiFi APs. An example of the measured RSSI vectors is presented in Fig. 2.7, where the first $N = 1000$ data were observed by Alice, while the following 1000 vectors were reported by Bob. Clearly, the RSSI vectors variant over time.

2.4.4.1 Key Generation Performance and Range Control

We use two criteria to evaluate the performance of the session key establishment: (1) the key generation rate that is the speed for Alice to generate $\mathbf{K}_A$ in bits per second, and (2) the key disagreement rate defined as the percentage of bits in Alice's key ($\mathbf{K}_A$) that are different from Bob's ($\mathbf{K}_B$).

Figure 2.8 provides the performance of Algorithm 2.2 in Experiment 1, with the time rounding parameter $\Upsilon = 1$, 2 and 3. It is shown in Fig. 2.8 that $\Upsilon = 2$ achieves both a high key generation rate and low key mismatching rate for all 17 scenarios. For instance, the lowest key generation rate is about 100 bps and the key disagreement rate is no more than 4%, if the Alice-Bob distance ranges between 1 and 55 m. With such a low error rate, the key disagreement can be conveniently addressed by the error correction codes such as BCH.

Next, as shown in Fig. 2.8b, the key generation rate decreases smoothly and slowly with the Alice-client distance. For instance, the key generation rate is above 100 bps even when Bob is about 50 m away from Alice and the key disagreement rate is less than 4%. Therefore, the key generation rate of Algorithm 2.2 can be used by Alice to determine whether Bob is in her proximity.

The maximum proximity range of the authentication based on the packet arrival time is much larger than that of the RSSI-based strategies. For example, Alice can authenticate clients as far as 50 m away by comparing the key generation rate of Algorithm 2.2 with the threshold Ξ. The parameter settings in Experiment 1 are listed in Table 2.4, with proximity range changing from 3 to 50 m. The system parameters, Δ and Ξ, are chosen according to the specified proximity range in the experiment.

Compared with most existing work, the proposed strategy provides a much larger maximum proximity range than most existing work. More specifically, considering 2.4 GHz WiFi ambient signals, the maximum proximity range of this strategy is around 50 m, while the maximum proximity ranges supported by ProxiMate in [15] and Ensemble in [14] are only 6.25 cm and 2 m, respectively. Moreover, this scheme provides higher key generation rates than ProxiMate. For example, in typical indoor environments, the key generation rate of this scheme, which is around 200 bps, is much higher than the 13 bps rate of ProxiMate in [15]. In addition, this scheme also provides more accurate authentication than Ensemble. For example, as shown in Section VI, this scheme has a small false rejection rate for clients within 3 m from Alice and false acceptance rate for clients more than 10 m away, which outperforms the 0.19 false rejection rate of Ensemble [32].

2.4.4.2   Room-Based Proximity Test

Experiment 2 contained six scenarios, with topology illustrated in Fig. 2.5a. In this experiment, Alice performed Algorithm 2.1 to decide whether Bob is in the same office. The performance of the room-based proximity test is presented in Fig. 2.9b, showing that the error rates for Alice to find a same-room client are mostly below 15%. We have also found that the ambient packet matching ratio is mostly above 40% when Alice and Bob are in the same room, or above 25% when they are in

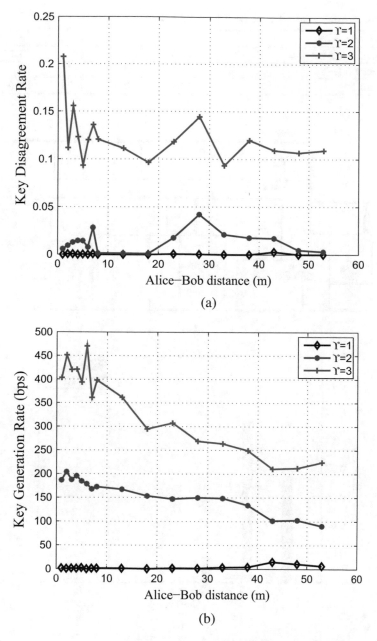

Fig. 2.8 Performance of the key generation algorithm (Algorithm ? ?), whose locations are shown in Fig. 2.5, with $\Upsilon = 1$, 2 and 3 (rounding to 0.1 s, 0.01 s and 1 ms). (a) Key disagreement rate between $\mathbf{K}_A$ and $\mathbf{K}_B$. (b) Key generation rate of $\mathbf{K}_A$

Table 2.4 Proximity control in the proposed authentication method in Experiment 1

Proximity range (m)	3	4	5	6
Threshold Δ in Algorithm 2.1 ($\Theta = 7.5$)	0.9	0.5	0.1	0.05
Proximity range (m)	10	20	40	50
Threshold Ξ	160	150	125	100

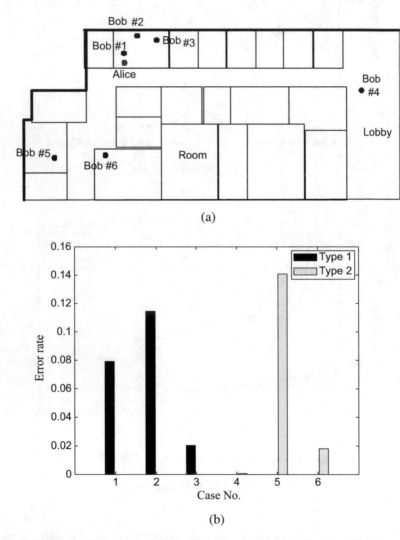

(a)

(b)

Fig. 2.9 Performance of the proximity-based authentication in Experiment 2. (**a**) Client placements in Virginia Tech Northern Virginia Center. (**b**) Error rates of the proximity test with $\Theta = 7.5$

different rooms. The results indicate that both clients have plenty of shared ambient packets to build the session key. Finally, we can see that the lowest session key generation rate is approximately as high as 248 bps. More details are given in [45].

Finally, we note that this work cannot achieve zero error rates, just like the other PHY-layer security schemes due to the properties of radio propagation. However, it can be used to enhance the security of LBS in wireless networks. For example, the proposed strategy provides a lightweight security protection for the LBS applications that do not require zero error rates in a wireless network without any pre-shared secret, trusted authority or public key infrastructure. On the other hand, for the applications with strict security requirements, the proposed scheme can serve as the bootstrap for the establishment of secure connections among the clients in the proximity and be incorporated with existing traditional security methods to achieve "100% security".

We have proposed a proximity-based authentication and key establishment scheme by exploiting the physical-layer features of ambient radio signals for LBS services in wireless networks, without requiring any pre-shared secret. Flexible range control is achieved by selecting the appropriate radio sources, such as ambient WiFi access points (APs), bluetooth devices and FM radios and choosing their suitable physical-layer features.

The system applies the Markov chain Monte Carlo implementation of the infinite Gaussian mixture model (IGMM) to classify the RSSIs of multiple ambient signals and thus determines whether a client is in the proximity. In the key establishment, clients generate session keys based on the normalized arrival time of their shared ambient packets.

The system does not disclose the client locations, and is robust against eavesdropping, spoofing, replay attacks and man-in-the-middle attacks outside the proximity. Experiments using laptops with WiFi packet analyzers in typical indoor environments have verified the efficacy of the security technique. By applying the IGMM model, the authentication is more accurate and is less sensitive to the radio propagation pattern than existing RSS and CIR-based authentication strategies. The key generation rate that can be as high as 248 bps in ideal cases is much higher than that of the RSS-based strategies. In the future, we will study how to incorporate this PHY-layer security strategy with the existing traditional security protocols to address the man-in-the-middle attacks inside the proximity.

2.5   RL Based PHY-Layer Authentication for VANETs Algorithm

We propose a PHY-layer rogue edge based authentication scheme for VANETs, which exploits the physical properties of the ambient radio signals received by both the mobile device and the edge node in a vehicle. More specifically, the mobile device and the edge node in the same vehicle have the same location tracing and thus can receive similar ambient radio signals from the source nodes such as the APs, BSs, RSUs and OBUs along the road. On the contrary, a rogue edge node Eve usually stays outside the target vehicle, and thus receive different ambient radio signals during the given time duration.

This scheme requires the edge node under test to send the PHY-layer feature such as the RSSIs of the ambient radio signals for a specific time duration and compares it with the record of the mobile device. Spoofing alarm is sent if the received PHY-layer feature trace is very different from the record of the mobile device at that time.

Let $I^{(k)} \in [0, 1]$ denote the importance of the computation task of the mobile device at time slot k. If the computation task is very important, the mobile device chooses a more advanced authentication scheme to detect rogue edge nodes. The false alarm rate at time slot k, denoted by $P_f^{(k)}$, is calculated, which is the ratio of the legitimate packets rejected by the PHY-layer authentication scheme by mistake. The mobile device evaluates the miss detection rate, denoted by $P_m^{(k)}$, which is the ratio of the spoofing packets falsely accepted at time k. The mobile device also estimated the ratio of the spoofing signals in each time slot denoted by $y^{(k)} \in [0, 1]$.

In the dynamic VANET authentication game, the mobile device chooses the spoofing detection mode and the test threshold based on the current state denoted by s^k. As the basis of the authentication, the state consists of the previous detection accuracy, the estimated spoofing probability in the last time slot, and the importance of the computation task of the mobile device in current time slot, i.e., $\mathbf{s}^{(k)} = [P_f^{(k-1)}, P_m^{(k-1)}, y^{(k-1)}, I^{(k)}]$.

As the next state observed by the mobile device is independent of the previous states and actions, for a given current state and authentication policy, the rogue edge detection process in the dynamic game can be formulated as an MDP. Therefore, the mobile device can apply reinforcement learning techniques such as Q-learning to derive the optimal strategy via trials without the knowledge of the VANET network model and the spoofing model.

The mobile device chooses the spoofing detection mode $x^{(k)} \in \mathbf{A}_0 = \{1, 2\}$ with a cost $C_f^{(k)}(x)$ and the mobile device also chooses the test threshold $\theta^{(k)}$ that is quantized into $X + 1$ levels, i.e., $\theta^{(k)} \in \mathbf{A}_1 = \{l/X\}_{0 \le l \le X}$. For convenience, we define $\mathbf{x}^{(k)} = \{x^{(k)}, \theta^{(k)}\} \in \mathbf{A} = \{\mathbf{A}_0, \mathbf{A}_1\}$, where $\mathbf{A}$ is the action set of the mobile device. The time index k in the superscript is omitted unless necessary. The spoofing detection mode cost is denoted by $C_f^{(k)}(x)$, which increases with x. For simplicity, we define $C_f^{(k)}(x) = x^2$.

Since the detection error rates depend on the test threshold, the mobile device has to select a proper value for $\theta^{(k)}$ in the detection. The utility of the mobile device at time slot k denoted by $u^{(k)}$ depends on the detection accuracy and the detection cost as follows,

$$u^{(k)}(\mathbf{x}, y) = -y^{(k)} I^{(k)} \left(\alpha_f P_f^{(k)} + \alpha_m P_m^{(k)} \right) - \beta C_f^{(k)} \tag{2.25}$$

where α_f (or α_m) is the weight factor of the false alarm (or the miss detection) in the detection, and $\beta \in [0, 1]$ is the cost of the spoofing detection mode cost.

Let $\mathbf{R}^{(k)}$ denote the trace vector under detection at time slot k. Similarly, we define the record trace vector of Bob at time slot k as $\hat{\mathbf{R}}$. The proposed rogue edge detection scheme provides two spoofing detection modes to detect whether the packet that Bob receives is indeed sent by Alice. Let $x = \{1, 2\}$ denote the spoofing detection modes, where $x = 1$ represents that Bob applies the historical RSSI records of the transmitter, i.e., $\mathbf{R}^{(k)} = r^{(k)}(i)$ and $\hat{\mathbf{R}} = \hat{r}(i)$, and $x = 2$ represents that Bob applies the transmitter historical RSSI records and the features records of the shared ambient radio signals, i.e., $\mathbf{R}^{(k)} = [r^{(k)}(i), \mathbf{f}^{(k)}]$, $\hat{\mathbf{R}} = [\hat{r}(i), \hat{\mathbf{f}}^{(k)}]$.

In the spoofing detection, Bob asks the edge node under detection to provide the PHY-layer feature traces. Once receiving the authentication message of the edge node under test, Bob authenticates the edge under test with the calculating physical information. If the trace vector based on the RSSI and the PHY-layer properties of the shared ambient radio signals, $\mathbf{R}^{(k)}$, is similar to the record trace of Bob, the authentication message is sent by Alice, or is sent by Eve otherwise. Bob calculates the test statistic of the hypothesis test denoted by Δ, and given by

$$\Delta = \frac{\left\| \mathbf{R}^{(k)} - \hat{\mathbf{R}} \right\|^2}{\left\| \hat{\mathbf{R}} \right\|^2} \tag{2.26}$$

where $\|\cdot\|$ is the Frobenius norm.

The test statistic is compared with the test threshold denoted by $\theta^{(k)} \geq 0$. If $\Delta < \theta^{(k)}$, the mobile device accepts the edge node under test. Otherwise, if $\Delta \geq \theta^{(k)}$, a spoofing alarm is sent. The mobile device applies higher-layer authentication techniques if the packet passes the PHY-layer authentication. If the i-th packet under test is accepted by the higher-layer authentication, the mobile device updates the RSSI record $\hat{r}(i)$ with $\hat{r}(i) \leftarrow r^{(k)}(i)$.

The rogue edge detection with Q-learning is based on the learning rate, denoted by $\alpha \in (0, 1]$, which shows the weight of the current experience. The discount factor $\delta \in [0, 1]$ corresponds to the uncertainty on the future utility. The Q-function of the action vector $\mathbf{x}^{(k)}$ at state $\mathbf{s}^{(k)}$ is denoted by $Q\left(\mathbf{s}^{(k)}, \mathbf{x}^{(k)}\right)$, and is updated according to iterative Bellman equation as follows:

$$Q\left(\mathbf{s}, \mathbf{x}\right) \leftarrow (1 - \alpha)Q\left(\mathbf{s}, \mathbf{x}\right)$$
$$+ \alpha \left(u\left(\mathbf{s}, \mathbf{x}\right) + \delta V\left(\mathbf{s}'\right) \right) \tag{2.27}$$

$$V\left(\mathbf{s}\right) = \max_{x \in A_0, \theta \in A_1} Q\left(\mathbf{s}, \mathbf{x}\right) \tag{2.28}$$

where $V\left(\mathbf{s}^{(k)}\right)$ maximizes the Q value over the rogue detection scheme.

Algorithm 2.4 Rogue detection with Q-learning

1: Initialize α, δ, ε, $P_f^{(0)}$, $P_m^{(0)}$, $y^{(0)}$, N, M, $\mathbf{A}$ and $\mathbf{s}^{(0)}$.
2: $\mathbf{Q} = 0$, $\mathbf{V} = 0$
3: **for** $k = 1, 2, \cdots$ **do**
4:      Evaluate the current data importance $I^{(k)}$
5:      $\mathbf{s}^{(k)} = [P_f^{(k-1)}, P_m^{(k-1)}, y^{(k-1)}, I^{(k)}]$
6:      Choose the spoofing detection mode $x^{(k)}$ and the test threshold $\theta^{(k)}$ via (2.29)
7:      Broadcast the time duration and the features of monitoring the ambient radio signals
8:      Store the feature information of the ambient radio packets as $\mathbf{f}^{(k)}$
9:      Receive the authentication message from the edge node under test
10:      Obtain $r^{(k)}(i)$ and $\mathbf{f}^{(k)}$ from the edge node under test
11:      **for** $i = 1, 2, \cdots, N$ **do**
12:        **if** $x^{(k)} = 1$ **then**
13:          $\mathbf{R}^{(k)} = r^{(k)}(i)$, $\hat{\mathbf{R}} = \hat{r}(i)$
14:        **else**
15:          $\mathbf{R}^{(k)} = [r^{(k)}(i), \mathbf{f}^{(k)}]$, $\hat{\mathbf{R}} = [\hat{r}(i), \hat{\mathbf{f}}^{(k)}]$
16:        **end if**
17:        Calculate Δ via (2.26)
18:        **if** $\Delta \leq \theta^{(k)}$ **then**
19:          Accept the i-th authentication message
20:          $\hat{r}(i) \leftarrow r^{(k)}(i)$
21:        **else**
22:          Send spoofing alarm
23:        **end if**
24:      **end for**
25:      Calculate $P_f^{(k)}$ and $P_m^{(k)}$
26:      Estimate the spoofing rate $y^{(k)}$ according to the detected illegal packets
27:      Evaluate $u^{(k)}$ via (2.25)
28:      Update $Q\left(\mathbf{s}^{(k)}, \mathbf{x}^{(k)}\right)$ via (2.27) and (2.28)
29: **end for**

The mobile device applies the ε-greedy policy to select the optimal action vector that maximizes the utility with a high probability $1 - \varepsilon$, and chooses the suboptimal action vectors with a small probability ε, i.e.,

$$\Pr\left(\mathbf{x}^*\right) = \begin{cases} 1 - \varepsilon, & \mathbf{x}^* = \arg \max_{x \in \mathbf{A}_0, \theta \in \mathbf{A}_1} Q\left(\mathbf{s}, \mathbf{x}\right) \\ \varepsilon/X, & \text{o.w.} \end{cases} \tag{2.29}$$

The PHY-layer rogue edge node detection with Q-learning is summarized in Algorithm 2.4.

Upon the chosen detection mode and test threshold $\mathbf{x}^{(k)}$, the mobile device calculates the test statistic Δ according to Eq. (2.26) and compares Δ with $\theta^{(k)}$ to determine whether the authentication message at time slot k is indeed sent by Alice. If Δ is less than the test threshold $\theta^{(k)}$, the mobile device updates the trace vector record and estimates the corresponding false alarm rate at time slot k. Otherwise, the mobile device sends a spoofing alarm, and estimates the miss detection rate at time slot k. For convenience, we define QARAS as the abbreviation of our proposed scheme.

2.6 Performance Evaluation

Simulations have been performed to evaluate the rogue edge detection scheme in a VANET with the VANET channel model given by Eq. (2.1) and initial network topology as shown in Fig. 2.10. In the simulations, we set the pass loss exponent as $n_0 = 3$, the reference distance as $d_0 = 10$ m, the weight factors between the false alarm rate and the miss detection rate as $\alpha_f = 0.5$, $\alpha_m = 1$, the cost of the spoofing detection mode cost as $\beta = 0.01$, and 5 ambient packets. Bob and Alice moved along the road at the same speed $v_1 = 10$ m/s.

As shown in Fig. 2.11a, the miss detection rate of QARAS decreases with time, from 16.3% at the beginning of the game to 0.26% after 2000 time slots, which is about 90.9% and 86.2% less than the QAR scheme developed in [4] and the QAAS scheme developed in [9], respectively. As shown in Fig. 2.11b, this scheme decreases the false alarm rate by 13.1% to 2.3% after 2000 time slots, which is 50.6% and 74.3% less than the benchmark QAR and QAAS, respectively.

As shown in Fig. 2.11c, the utility of the mobile device increases with time, and this scheme increases the utility than the benchmark schemes. For example, the utility of the mobile device increases by about 65.7% after 2000 time slots at convergence, which is about 39.1% and 73.9% higher than QAR and QAAS, respectively. In addition, the QARAS scheme has a faster learning speed, e.g.,

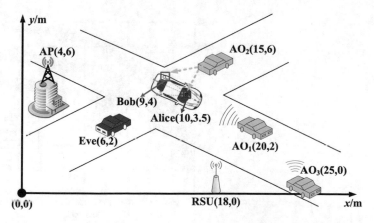

Fig. 2.10 Network topology in the simulation setting in meters, in which Bob, Alice and Eve extract and store the PHY-layer feature information of the ambient radio signals during the specific time duration to detect rogue edge Eve in a VANET with 5 ambient radio sources, including the AP, RSU, and three OBUs

Fig. 2.11 Performance of the PHY-layer rogue edge detection in the VANET as shown in Fig. 2.2, with the path loss exponent $n_0 = 3$, the reference distance $d_0 = 10$ m, the weight factors between the false alarm rate and the miss detection rate $\alpha_f = 0.5$, $\alpha_m = 1$, the cost of the spoofing detection mode cost $\beta = 0.01$, and 5 ambient packets. (**a**) Miss detection rate. (**b**) False alarm rate. (**c**) Utility of Bob

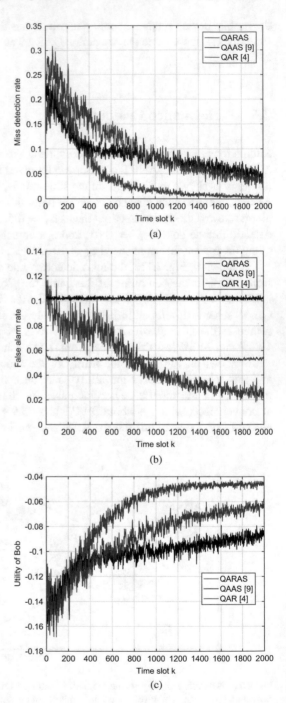

this scheme takes about 1200 time slots (about 2.65 s) to achieve the optimal authentication policy, while it will take more time slots to converge for QAR and QAAS.

2.7 Summary

In this chapter, we have presented a PHY-layer authentication scheme to detect rogue edge nodes for VANETs. This scheme depends on the physical properties of the ambient radio signals received by both the mobile device and the serving edge node and uses Q-learning to enable a mobile device to achieve the optimal authentication policy in the dynamic VANET without knowing of the VANET model or the attack model. We have also proposed a proximity-based authentication and key establishment scheme by exploiting the physical-layer features of ambient radio signals for LBS services in wireless networks, without requiring any pre-shared secret. Flexible range control is achieved by selecting the appropriate radio sources, such as ambient WiFi access points (APs), bluetooth devices and FM radios and choosing their suitable physical-layer features. Simulation results show that our proposed scheme exceeds the benchmark schemes such as QAR and QAAS with higher detection accuracy and higher utility. For instance, the miss detection rate and the false alarm rate of this scheme are 0.26% and 2.3% after 2000 time slots, which are 86.2% and 74.3% less than the benchmark strategy QAAS, respectively.

References

1. K. Zeng, K. Govindan, and P. Mohapatra, "Non-cryptographic authentication and identification in wireless networks," *IEEE Wireless Commun.*, vol. 17, no. 5, pp. 56–62, 2010.
2. K. Zaidi, M. B. Milojevic, V. Rakocevic, A. Nallanathan, and M. Rajarajan, "Host-based intrusion detection for VANETs: A statistical approach to rogue node detection," *IEEE Trans. Vehicular Technology*, vol. 65, no. 8, pp. 6703–6714, Aug. 2016.
3. S. Basudan, X. Lin, and K. Sankaranarayanan, "A privacy-preserving vehicular crowdsensing-based road surface condition monitoring system using fog computing," *IEEE Internet of Things Journal*, vol. 4, no. 3, pp. 772–782, Jun. 2017.
4. L. Xiao, Y. Li, G. Han, G. Liu, and W. Zhuang, "PHY-layer spoofing detection with reinforcement learning in wireless networks," *IEEE Trans. Vehicular Technology*, vol. 65, no. 12, pp. 10037–10047, Dec. 2016.
5. C. Pei, N. Zhang, X. Shen, and J. W. Mark, "Channel-based physical layer authentication," in *Proc. IEEE Global Commun. Conf.*, pp. 4114–4119, Austin, TX, Dec. 2014.
6. N. Wang, T. Jiang, S. Lv, and L. Xiao, "Physical-layer authentication based on extreme learning machine," *IEEE Commun. Letters*, vol. 35, no. 3, pp. 1089–7798, Jul. 2017.
7. L. Shi, M. Li, S. Yu, and J. Yuan, "BANA: Body area network authentication exploiting channel characteristics," *IEEE J. Sel. Areas Commun.*, vol. 31, no. 9, pp. 1803–1816, Sept. 2013.
8. L. Xiao, Q. Yan, W. Lou, G. Chen, and Y. T. Hou, "Proximity-based security techniques for mobile users in wireless networks," *IEEE Trans. Inf. Forensics Security*, vol. 8, no. 12, pp. 2089–2100, Oct. 2013.

9. J. Liu, L. Xiao, G. Liu, and Y. Zhao, "Active authentication with reinforcement learning based on ambient radio signals," *Springer Multimedia Tools and Applications*, vol. 76, no. 3, pp. 3979–3998, Oct. 2015.

10. A. Wasef, Y. Jiang, and X. Shen, "DCS: An efficient distributed-certificate-service scheme for vehicular networks," *IEEE Trans. Vehicular Technology*, vol. 59, no. 2, pp. 533–549, Jul. 2009.

11. F. J. Liu, X. Wang, and S. L. Primak, "A two dimensional quantization algorithm for CIR-based physical layer authentication," in *Proc. IEEE Int'l Conf. Commun. (ICC)*, pp. 4724–4728, Budapest, Jun. 2013.

12. X. Du, D. Shan, K. Zeng, and L. Huie, "Physical layer challenge-response authentication in wireless networks with relay," in *Proc. IEEE Int'l Conf. Computer Commun. (INFOCOM)*, pp. 1276–1284, Toronto, ON, Apr. 2014.

13. C. Wang, X. Zheng, Y. Chen, and J. Yang, "Locating rogue access point using fine-grained channel information," *IEEE Trans. Mobile Computing*, vol. 16, no. 9, pp. 2560–2573, Sept. 2017.

14. S. Mathur, R. Miller, A. Varshavsky, W. Trappe, and N. Mandayam, "Proximate: proximity-based secure pairing using ambient wireless signals," in *Proc. ACM Int'l Conf Mobile systems, applications, and services*, pp. 211–224, Jun. 2011.

15. H. Han, F. Xu, C. C. Tan, Y. Zhang, and Q. Li, "VR-defender: Self-defense against vehicular rogue APs for drive-thru internet," *IEEE Trans. Vehicular Technology*, vol. 63, no. 8, pp. 3927–3934, Oct. 2014.

16. K. Zaidi, M. B. Milojevic, V. Rakocevic, A. Nallanathan, and M. Rajarajan, "Host-based intrusion detection for VANETs: A statistical approach to rogue node detection," *IEEE Trans. Vehicular Technology*, vol. 65, no. 8, pp. 6703–6714, Aug. 2016.

17. A. Abdallah and X. Shen, "Lightweight authentication and privacy-preserving scheme for V2G connections," *IEEE Trans. Vehicular Technology*, vol. 66, no. 3, pp. 2615–2629, Mar. 2017.

18. X. Wan, L. Xiao, Q. Li, and Z. Han, "PHY-layer authentication with multiple landmarks with reduced communication overhead," in *Proc. IEEE Int'l Conf. Commun. (ICC)*, Paris, France, May 2017.

19. M. Li, W. Lou, and K. Ren, "Data security and privacy in wireless body area networks," *IEEE Wireless Communications*, vol. 17, pp. 51–58, February 2010.

20. X. Liang, R. Lu, C. Le, X. Lin, and X. Shen, "Pec: A privacy-preserving emergency call scheme for mobile healthcare social networks," *Journal of Communications and Networks*, vol. 13, pp. 102–112, April 2011.

21. A. Narayanan and V. Shmatikov, "De-anonymizing social networks," in *Proc. IEEE Symposium on Security and Privacy*, 2009.

22. J. Tsai, P. Kelley, L. Cranor, and N. Sadeh, "Location-sharing technologies: Privacy risks and controls," *ISJLP*, vol. 6, pp. 119–317, August 2009.

23. G. Ghinita, P. Kalnis, A. Khoshgozaran, C. Shahabi, and K. Tan, "Private queries in location based services: anonymizers are not necessary," in *Proc. ACM SIGMOD international conference on Management of data*, 2008.

24. W. He, X. Liu, and M. Ren, "Location cheating: A security challenge to location-based social network services," in *IEEE ICDCS*, 2011.

25. W. Chang, J. Wu, and C. Tan, "Enhancing mobile social network privacy," in *Proc. IEEE Globecom*, 2011.

26. Z. Zhu and G. Cao, "Applaus: A privacy-preserving location proof updating system for location-based services," in *Proc. IEEE INFOCOM*, 2011.

27. L. Siksnys, J. Thomsen, S. Saltenis, M. Yiu, and O. Andersen, "A location privacy aware friend locator," *Advances in Spatial and Temporal Databases*, vol. 5644, pp. 405–410, 2009.

28. A. Narayanan, N. Thiagarajan, M. Lakhani, M. Hamburg, and D. Boneh., "Location privacy via private proximity testing," in *Proc. Network and Distributed System Security Symposium (NDSS)*, 2011.

29. N. Talukder and S. Ahamed, "Preventing multi-query attack in location-based services," in *Proc. ACM conference on Wireless network security*, 2010.

30. R. Mayrhofer and H. Gellersen, "Shake well before use: intuitive and secure pairing of mobile devices," *IEEE Trans. Mobile Computing*, vol. 8, pp. 792–806, June 2009.
31. A. Varshavsky, A. Scannell, A. LaMarca, and E. Lara, "Amigo: Proximity-based authentication of mobile devices," in *Proc. UbiComp*, 2007.
32. A. Kalamandeen, A. Scannell, E. de Lara, A. Sheth, and A. LaMarca, "Ensemble: cooperative proximity-based authentication," in *Proc. ACM 8th international conference on Mobile systems, applications, and services*, 2010.
33. S. Mathur, R. Miller, A. Varshavsky, and W. Trappe, "Proximate: Proximity-based secure pairing using ambient wireless signals," in *Proc. ACM MobySys*, 2011.
34. Y. Zheng, M. Li, W. Lou, and T. Hou, "Sharp: Private proximity test and secure handshake with cheat-proof location tags," in *Proc. European Symposium on Research in Computer Security (ESORICS)*, 2012.
35. S. Eberz, M. Strohmeier, M. Wilhelm, and I. Martinovic, "A practical man-in-the-middle attack on signal-based key generation protocols," in *Proc. 17th European Symposium on Research in Computer Security (ESORICS)*, 2012.
36. A. Goldsmith, *Wireless Communications*. chapter 3, Cambridge University Press, 2005.
37. C. Rasmussen, "The infinite gaussian mixture model," *Advances in neural information processing systems*, pp. 554–560, 2000.
38. N. Nguyen, G. Zheng, Z. Han, and R. Zheng, "Device fingerprinting to enhance wireless security using nonparametric bayesian method," in *Proc. IEEE INFOCOM*, 2011.
39. N. Nguyen, R. Zheng, and Z. Han, "On identifying primary user emulation attacks in cognitive radio systems using nonparametric bayesian classification," *IEEE Trans. Signal Processing*, vol. 60, pp. 1432–1445, March 2012.
40. C. Bishop, *Pattern recognition and machine learning*. Springer Press, 2006.
41. Z. Lin, D. Kune, and N. Hopper, "Efficient private proximity testing with gsm location sketches," *Financial Cryptography and Data Security*, pp. 73–88, 2012.
42. S. Mascetti, C. Bettini, D. Freni, X. Wang, and S. Jajodia, "Privacy-aware proximity based services," in *Proc. International Conference on Mobile Data Management: Systems, Services and Middleware*, 2009.
43. J. Meyerowitz and R. R. Choudhury, "Hiding stars with fireworks: location privacy through camouflage," in *Proc. international conference on Mobile computing and networking*, 2009.
44. L. Siksnys, J. Thomsen, S. Saltenis, and M. Yiu, "Private and flexible proximity detection in mobile social networks," in *Proc. International Conference on Mobile Data Management*, 2010.
45. L. Xiao, Q. Yan, W. Lou, and T. Hou, "Proximity-based security using ambient radio signals," in *Proc. IEEE ICC*, 2013, to appear.
46. B. Azimi, A. Kiayias, A. Mercado, and B. Yener, "Robust key generation from signal envelopes in wireless networks," in *Proc. ACM Conference on Computer and Communications Security*, 2007.
47. C. Ye, S. Mathur, A. Reznik, Y. Shah, W. Trappe, and N. Mandayam, "Information-theoretically secret key generation for fading wireless channels," *IEEE Trans. Information Forensics and Security*, vol. 5, pp. 240–254, 2010.
48. T. Aono, K. Higuchi, T. Ohira, B. Komiyama, and H. Sasaoka, "Wireless secret key generation exploiting reactance-domain scalar response of multipath fading channels," *IEEE Transactions on Antennas and Propagation*, vol. 53, pp. 3776–3784, Nov. 2005.
49. J. Croft, N. Patwari, and S. Kasera, "Robust uncorrelated bit extraction methodologies for wireless sensors," in *Proc. ACM/IEEE International Conference on Information Processing in Sensor Networks (IPSN)*, 2010.
50. S. Mathur, W. Trappe, N. Mandayam, C. Ye, and A. Reznik, "Radio-telepathy. Extracting a secret key from an unauthenticated wireless channel," in *Proc. ACM 14th annual conference on mobile computing and systems (MobiCom 2008)*, 2008.
51. Q. Wang, H. Su, K. Ren, and K. Kim, "Fast and scalable secret key generation exploiting channel phase randomness in wireless networks," in *Proc. IEEE INFOCOM*, 2011.

Chapter 3
Learning While Offloading: Task Offloading in Vehicular Edge Computing Network

In vehicular edge computing (VEC) systems, vehicles can contribute their computing resources to the network, and help other vehicles or pedestrians to process their computation tasks. However, the high mobility of vehicles leads to a dynamic and uncertain vehicular environment, where the network topologies, channel states and computing workloads vary fast across time. Therefore, it is challenging to design task offloading algorithms to optimize the delay performance of tasks. In this chapter, we consider the task offloading among vehicles, and design learning-based task offloading algorithms based on the multi-armed bandit (MAB) theory, which enable vehicles to learn the delay performance of their surrounding vehicles while offloading tasks. We start from the single offloading case where each task is offloaded to one vehicle to be processed, and propose an adaptive learning-based task offloading (ALTO) algorithm, by jointly considering the variations of surrounding vehicles and the input data size. To further improve the reliability of the computing services, we introduce the task replication technique, where the replicas of each task is offloaded to multiple vehicles and processed by them simultaneously, and propose a learning-based task replication algorithm (LTRA) based on combinatorial MAB. We prove that the proposed ALTO and LTRA algorithms have bounded learning regret, compared with the genie-aided optimal solution. And we also build a system level simulation platform to evaluate the proposed algorithms in the realistic vehicular environment.

3.1 Vehicular Edge Computing Architecture

In Vehicular Edge Computing (VEC) systems, some vehicles and infrastructures like road side units (RSUs) provide their computing resources to the network, while computation tasks are generated by other vehicles and pedestrians from various kinds of applications [1–4]. The development of vehicle-to-everything (V2X)

© Springer Nature Switzerland AG 2019

L. Xiao et al., *Learning-based VANET Communication and Security Techniques*, Wireless Networks, https://doi.org/10.1007/978-3-030-01731-6_3

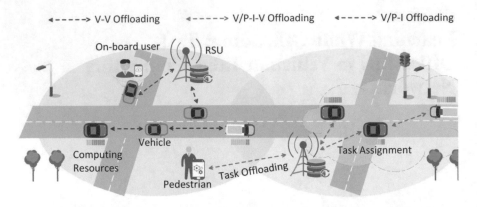

Fig. 3.1 Architecture of the VEC system and three major offloading modes

protocols enable different kinds of communications in the VANETs, including vehicle-to-vehicle (V2V), vehicle-to-infrastructure (V2I) and vehicle-to-pedestrian (V2P) communications [5]. Thus computation tasks can be offloaded to the VEC system via diverse routes. The architecture of the VEC system is illustrated in Fig. 3.1. Based on the task collector and executor, we classify the task offloading into three major modes.

- *Vehicle-Vehicle (V-V) Offloading*: Vehicles offload their computation tasks directly to the surrounding vehicles who have sufficient computing resources. In this mode, offloading decisions are often made by each vehicle in a distributed way. Challenges include that, vehicles may not obtain the global state information of surrounding vehicles, including wireless channel states and available computing resources. And it is difficult for vehicles to cooperate with each other.
- *Pedestrian/Vehicle-Infrastructure-Vehicle (P/V-I-V) Offloading*: There might be no neighboring vehicles for task offloading, or that the surrounding vehicles can not satisfy the computation requirements of tasks. In these cases, the roadside infrastructures can help to collect tasks from pedestrians or vehicles, and assign them to other vehicles in a centralized manner. Result feedbacks are also collected by the infrastructures and finally transmitted back to the task generators. Infrastructures can usually get global state information to optimize the task scheduling, but the signaling overhead is much higher than the V-V offloading.
- *Pedestrian/Vehicle-Infrastructure (P/V-I) Offloading*: Tasks from pedestrians or vehicles are offloaded to the infrastructures and processed by them. Although infrastructures can optimally allocate radio and computing resources to the tasks, the limited amount of resource can not always satisfy the task requirements.

In this chapter, we mainly focus on the V-V offloading mode, and design distributed task offloading algorithms in the following sections.

3.2   Vehicle-to-Vehicle Task Offloading: Procedure and Problem Formulation

We focus on the V-V offloading in the VEC system, and classify the vehicles into two categories: *task vehicles (TaVs)* are the vehicles that require computation task offloading, while *service vehicles (SeVs)* are the vehicles that can provide their surplus computing resources to the network. Note that each vehicle is not fixed as a TaV or SeV during the trip, which depends on whether it requires computing services, and whether it has sufficient and shareable computing resources [6].

TaVs can offload tasks to their surrounding SeVs. Each TaV discovers the SeVs within its communication range, and chooses those in the same moving direction as candidate SeVs. Here the moving direction, as well as vehicle ID, location and velocity of each neighboring SeV can be acquired by the TaV, through vehicular communication protocols such as beacons of dedicated short-range communication (DSRC) standards [7].

For each TaV, there might be multiple candidate SeVs, while each task is offloaded to one SeV and processed by it, without offloaded to other SeVs or RSUs furthermore. An illustration of V-V offloading is shown in Fig. 3.2, where TaV 1 regards SeVs 1–3 as candidates, and currently the task is offloaded to SeV 3.

We will design distributed task offloading algorithm in the following sections, where each TaV makes its own offloading decisions on which SeV should serve each task, in order to minimize the average offloading delay. Besides, we do not assume any prior information of the mobility models of vehicles, and do not consider the inter-TaV cooperations.

3.2.1   Task Offloading Procedure

Since we consider distributed offloading without inter-vehicle cooperations, we focus on a single TaV of interest and formulate the task offloading problem. Consider a discrete-time VEC system with a total number of T time periods. Within each time period, there are four offloading procedures as follows:

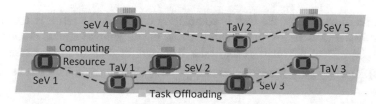

Fig. 3.2 An illustration of V-V offloading in the VEC system

SeV discovery: The TaV discovers candidate SeVs at the beginning of each time period. In time period t, the candidate SeV set is denoted by $\mathcal{N}(t)$. Note that $\mathcal{N}(t)$ changes across time due to the movements of vehicles. And since no prior information of vehicles is available, the TaV has no idea about when each SeV will appear and disappear. Assume that $\mathcal{N}(t) \neq \emptyset$ for $\forall t$, otherwise the TaV can offload tasks to RSUs, which is beyond the scope of this work.

Task upload: For the computation task generated in time period t, the TaV chooses one SeV $n \in \mathcal{N}(t)$ and uploads the data to be processed. The input data size of the task is denoted by x_t (in bits), the uplink wireless channel state between the TaV and each candidate SeV $n \in \mathcal{N}(t)$ is $h_{t,n}^{(u)}$, and the interference power at SeV n is denoted by $I_{t,n}^{(u)}$. Denote $r_{t,n}^{(u)}$ as the uplink transmission rate between the TaV and candidate SeV n, which can be given by

$$r_{t,n}^{(u)} = W \log_2 \left(1 + \frac{P h_{t,n}^{(u)}}{\sigma^2 + I_{t,n}^{(u)}} \right), \tag{3.1}$$

where P is the transmission power, W is the channel bandwidth, and σ^2 is the noise power. Then the uplink transmission delay $d_{\text{up}}(t, n)$ is given by

$$d_{\text{up}}(t, n) = \frac{x_t}{r_{t,n}^{(u)}}. \tag{3.2}$$

Task execution: After receiving the data from the TaV, the chosen SeV n starts to process the computation task. Denote the computation intensity of the task in time period t as w_t (in CPU cycles per bit), which represents the number of CPU cycles required to execute one bit input data. According to the linear computational complexity model, the total computing workload is $x_t w_t$ [8].

The maximum CPU frequency F_n is adopted to describe the computing capability of each candidate SeV n (in CPU cycles per bit). During each time period t, SeVs may need to process multiple tasks, and the CPU frequency allocated to the task of the TaV of interest is given by $f_{t,n}$. Then the computation delay can be written as

$$d_{\text{com}}(t, n) = \frac{x_t w_t}{f_{t,n}}. \tag{3.3}$$

Result feedback: After the task is completely executed, the result is transmitted back from the chosen SeV to the TaV. Denote the downlink wireless channel state by $h_{t,n}^{(d)}$, let $I_t^{(d)}$ be the interference at the TaV, and the data size of the result be y_t (in bits). The downlink transmission rate $r_{t,n}^{(d)}$ from SeV n to the TaV can be written as

$$r_{t,n}^{(d)} = W \log_2 \left(1 + \frac{P h_{t,n}^{(d)}}{\sigma^2 + I_t^{(d)}} \right). \tag{3.4}$$

And the downlink transmission delay is

$$d_{\text{dow}}(t, n) = \frac{y_t}{r_{t,n}^{(d)}}. \tag{3.5}$$

Finally, the sum offloading delay $d_{\text{sum}}(t, n)$, representing the total delay of offloading the task to SeV n at time t, can be given by

$$d_{\text{sum}}(t, n) = d_{\text{up}}(t, n) + d_{\text{com}}(t, n) + d_{\text{dow}}(t, n). \tag{3.6}$$

3.2.2 Problem Formulation

Our objective is to minimize the average offloading delay over T time periods, where each TaV decides which SeV should serve each task. The problem is formulated as

$$\textbf{P1:} \quad \min_{a_1,\dots,a_T} \frac{1}{T} \sum_{t=1}^{T} d_{\text{sum}}(t, a_t), \tag{3.7}$$

where $a_t \in \mathcal{N}(t)$ is denoted as the index of the chosen SeV in time period t.

Lack of global state information: In time period t, the properties of task, including the input and output data size x_t, y_t, as well as the computation intensity w_t is easy to be acquired by the TaV before offloading. However, other kinds of the state information, including the CPU frequency $f_{t,n}$, transmission rates $r_{t,n}^{(u)}$, $r_{t,n}^{(d)}$ and interference $I_{t,n}^{(u)}$ are owned by each candidate SeV n. Due to the movements of vehicles, and the shareable computing resources of SeVs, these parameters are difficult to model or to predict. Besides, if all candidate SeVs report their state information, the signaling overhead will increase drastically. Therefore, the TaV cannot acquire the global state information of SeVs, and cannot be aware of the optimal SeV with the lowest offloading delay for each task.

Learning while offloading: To deal with the challenge that the TaV lacks the global state information when making offloading decisions, our solution is *learning while offloading*. That is, the TaV can learn the delay performance of candidate SeVs by observing the delay of previously completed tasks. To be specific, the offloading decision a_t made in time period t is only based on the historical delay observations $d_{\text{sum}}(1, a_1), d_{\text{sum}}(2, a_2), \dots, d_{\text{sum}}(t-1, a_{t-1})$, rather than the current state information of all candidate SeVs.

We aim at designing a leaning-based task offloading algorithm, in order to minimize the expected average offloading delay, given by

$$\textbf{P2:} \quad \min_{a_1,\dots,a_T} \frac{1}{T} \mathbb{E}\left[\sum_{t=1}^{T} d_{\text{sum}}(t, a_t) \right]. \tag{3.8}$$

A simplification to problem **P2** is considered in the following part, by assuming that the output and input data size ratio y_t/x_t, as well as the computation intensity w_t remain identical over time. Only the input data volume x_t changes across time. In fact, this is a reasonable assumption when tasks are generated from the same kind of application. Let $y_t/x_t = \alpha_0$ and $w_t = \omega_0$ for $\forall t$. Then the sum offloading delay to SeV n can be written as

$$d_{\text{sum}}(t, n) = x_t \left(\frac{1}{r_{t,n}^{(u)}} + \frac{\alpha_0}{r_{t,n}^{(d)}} + \frac{\omega_0}{f_{t,n}} \right). \tag{3.9}$$

We define the *bit offloading delay* $u(t, n)$ as the total delay of offloading one bit of input data to SeV n at time t, which reflects the communication and computing service capability of candidate SeVs. The TaV has to learn the bit offloading delay of all the candidate SeVs, which can be given by

$$u(t, n) = \frac{1}{r_{t,n}^{(u)}} + \frac{\alpha_0}{r_{t,n}^{(d)}} + \frac{\omega_0}{f_{t,n}}. \tag{3.10}$$

And then problem **P2** can be transformed to

$$\textbf{P3:} \quad \min_{a_1,\dots,a_T} \frac{1}{T} \mathbb{E}\left[\sum_{t=1}^{T} x_t u(t, n) \right]. \tag{3.11}$$

3.3   Learning While Offloading Based on Multi-armed Bandit

In this section, we solve the problem **P3** by designing a learning-based task offloading algorithm based on the multi-armed bandit (MAB) theory, and characterize its performance bound.

3.3.1   Adaptive Learning-Based Task Offloading Algorithm

Our task offloading algorithm is designed based on the MAB theory [9]. In classical MAB problems, there is a player facing a fixed number of candidate actions. At each time, the player selects an action and gets a loss (or reward). The loss distribution of each action is unknown to the player in prior and needs to be learned. And the objective is to design learning algorithms that minimize the expected cumulative loss (or maximize expected cumulative reward). The main challenge is to balance the tradeoff between *exploration and exploitation* during the learning process: to explore different actions and acquire more information of their loss distributions, or to exploit the existing information and select the empirically best action to minimize the instantaneous loss. The classical MAB problem has been widely investigated, and algorithms base on upper confidence bound (UCB) have been proposed, providing strong performance guarantee [9].

In our task offloading problem, each candidate SeV corresponds to an action, and the TaV is the player. However, there are two new challenges in our problem: the time varying candidate SeV set $\mathcal{N}(t)$, and the diverse input data size x_t. To deal with these challenges, we propose an Adaptive Learning-based Task Offloading (ALTO) algorithm, by jointly considering the occurrence time of vehicles and the input data sizes of tasks, as shown in Algorithm 3.1. Parameter β is a constant weight factor, $k_{t,n}$ is the number of tasks offloaded to SeV n till time t, and t_n records the occurrence time of SeV n. Parameter $\tilde{x}_t$ is the normalized input data size within $[0, 1]$:

$$\tilde{x}_t = \max\left\{\min\left(\frac{x_t - x^-}{x^+ - x^-}, 1\right), 0\right\}, \qquad (3.12)$$

where x^+ and x^- are the upper and lower thresholds for normalization. Particularly, if $x^+ = x^-$, $\tilde{x}_t = 0$ when $x_t \leq x^-$, and $\tilde{x}_t = 1$ when $x_t > x^-$.

In Algorithm 3.1, Lines 3–5 correspond to the initialization phase, which is triggered whenever new SeV appears as candidate. The TaV offloads a task to the new SeV once, and obtain an initial evaluation of its service capability. Lines 7–12 represent the continuous learning phase. The utility function is shown in (3.13),

---

Algorithm 3.1 ALTO: Adaptive learning-based task offloading algorithm

1: Initialize $\alpha_0, \omega_0, \beta, x^+$ and x^-.
2: **for** $t = 1, \ldots, T$ **do**
3:     **if** Any SeV $n \in \mathcal{N}(t)$ has not be chosen by the TaV **then**
4:         Connect to SeV n once.
5:         Update $\bar{u}_{t,n} = d_{\text{sum}}(t, n)/x_t, k_{t,n} = 1, t_n = t$.
6:     **else**
7:         Observe x_t, calculate $\tilde{x}_t$.
8:         Calculate the utility function of each candidate SeV $n \in \mathcal{N}(t)$:

$$\hat{u}_{t,n} = \bar{u}_{t-1,n} - \sqrt{\frac{\beta(1 - \tilde{x}_t)\ln(t - t_n)}{k_{t-1,n}}}. \qquad (3.13)$$

9:         Offload the task to SeV a_t, such that:

$$a_t = \arg\min_{n \in \mathcal{N}(t)} \hat{u}_{t,n}. \qquad (3.14)$$

10:         Observe the sum offloading delay $d_{\text{sum}}(t, a_t)$.
11:         Update $\bar{u}_{t,a_t} \leftarrow \frac{\bar{u}_{t-1,a_t} k_{t-1,a_t} + d_{\text{sum}}(t,a_t)/x_t}{k_{t-1,a_t} + 1}$.
12:         Update $k_{t,a_t} \leftarrow k_{t-1,a_t} + 1$.
13:     **end if**
14: **end for**

---

which is inspired by the volatile MAB [10] and opportunistic MAB [11]. The utility function adds the empirical bit offloading delay $\bar{u}_{t,n}$ and a padding function (the latter term). Compared with the utility function of existing UCB algorithms, we jointly consider the occurrence time t_n of SeV and the normalized input data volume $\tilde{x}_t$, which can effectively balance the exploration and exploitation and enable the proposed ALTO algorithm to adapt to the dynamic vehicular task offloading environment. Then the TaV finds the SeV with minimum utility according to (3.14), with linear searching complexity $O(|\mathcal{N}(t)|)$, where $|\mathcal{N}(t)|$ is the number of candidate SeVs.

The key design ideas of the padding function in (3.13), namely *occurrence-awareness* and *input-awareness*, are further illustrated as follows.

Occurrence-awareness: ALTO algorithm can guide the learning process according to the occurrence time of SeVs. When SeV n just arrives, $\sqrt{\frac{\ln(t-t_n)}{k_{t-1,n}}}$ is large so that the TaV tends to explore this SeV to learn its service capability. When a SeV has been connected for many times, the padding function decreases, so that the TaV tends to exploit the empirical information.

Input-awareness: The normalized input data size $\tilde{x}_t$ brings a weighted factor to the padding function. When x_t is small, $\sqrt{1-\tilde{x}_t}$ is large, so that the TaV tends to explore more. Even if the service capability of the explored SeV is poor, the sum offloading delay is not too large due to the small input data volume. On the contrary, when x_t is large, $\sqrt{1-\tilde{x}_t}$ is small, and thus the TaV tends to exploit the learned information to avoid long offloading delay.

3.3.2 Performance Guarantee

To characterize the performance of the proposed ALTO algorithm, we adopt the performance metric *learning regret*, which is widely used in the MAB theory.

Define an *epoch* as the duration in which candidate SeVs remain identical. Let B be the total number of epochs, and $\mathcal{N}_b$ the candidate SeV set during the bth epoch, with $b = 1, 2, \ldots, B$. Let t_b and t'_b be the start and end time of the bth epoch. It is easy to see that $t_1 = 1$ and $t'_B = T$. To carry out theoretical analysis, we assume that the bit offloading delay $u(t, n)$ of each SeV n is i.i.d. over time and independent of others. We will show through simulations that without this assumption, the proposed ALTO algorithm still works well.

Let $\mu_n = \mathbb{E}_t[u(t, n)]$ be the mean bit offloading delay of candidate SeV n, $\mu_b^* = \min_{n \in \mathcal{N}_b} \mu_n$ the optimal mean bit offloading delay in the bth epoch, and $a_b^* = \arg \min_{n \in \mathcal{N}_b} \mu_n$. Note that for $\forall n$, μ_n is unknown to the TaV in prior.

The learning regret is the expected cumulative loss of sum offloading delay caused by learning, compared with the genie-aided optimal policy where the TaV can always offload tasks to the optimal SeV a_b^* during each epoch b. The learning regret by time period T is given by

$$R_T = \sum_{b=1}^{B} \mathbb{E}\left[\sum_{t=t_b}^{t_b'} x_t\left(u(t,n) - \mu_b^*\right)\right], \tag{3.15}$$

In the following, we analyze the impacts of the occurrence time t_n and normalized input data size $\tilde{x}_t$ on the learning regret separately.

3.3.2.1 Learning Regret Under Dynamic SeV Set and Identical Input

We then analyze the impact of random input data size on the learning regret, focusing on a single epoch. Let $B = 1$, and we omit the subscript b for simplicity. Let $a^* = \arg\min_{n \in \mathcal{N}_1} \mu_n$, and $\mu^* = \min_{n \in \mathcal{N}_1} \mu_n$.

The learning regret can be simplified as

$$R_T = \mathbb{E}\left[\sum_{t=1}^{T} x_t(u(t,n) - \mu^*)\right]. \tag{3.16}$$

The learning regret R_T under varying input data volume and fixed candidate SeV set is shown as follows.

Theorem 3.1 *Let $\beta_0 = 2$, and $\mathbb{P}\{x_t \leq x^-\} > 0$, we have:*

(1) *When $x^+ \geq x^-$, the expectation of the number of tasks $k_{T,n}$ offloaded to any SeV $n \neq a^*$ can be bounded as*

$$\mathbb{E}[k_{T,n}] \leq \frac{8 \ln T}{\delta_n^2} + O(1). \tag{3.17}$$

(2) *With $x^+ = x^-$, the learning regret can be bounded as*

$$R_T \leq u_m \sum_{n \neq a^*} \left[\frac{8 \ln T \, \mathbb{E}[x_t | x_t \leq x^-]}{\delta_n} + O(1)\right], \tag{3.18}$$

where $\mathbb{E}[x_t | x_t \leq x^-]$ is the expectation of x_t on the condition that $x_t \leq x^-$, $u_m = \sup_{t,n} u(t,n)$, and $\delta_n = (\mu_n - \mu^*)/u_m$.

Proof When $\beta_0 = 2$ and $B = 1$, the utility function (3.13) can be simplified as

$$\hat{u}_{t,n} = \bar{u}_{t-1,n} - u_m \sqrt{\frac{2(1 - \tilde{x}_t) \ln t}{k_{t-1,n}}}. \tag{3.19}$$

The decision making function in (3.14) can be written as

$$a_t = \arg\min_{n \in \mathcal{N}_1} \hat{u}_{t,n} = \arg\min_{n \in \mathcal{N}_1} \left\{ \bar{u}_{t-1,n} - u_m \sqrt{\frac{2(1-\tilde{x}_t)\ln t}{k_{t-1,n}}} \right\}$$

$$= \arg\min_{n \in \mathcal{N}_1} \left\{ \frac{\bar{u}_{t-1,n}}{u_m} - \sqrt{\frac{2(1-\tilde{x}_t)\ln t}{k_{t-1,n}}} \right\}$$

$$= \arg\max_{n \in \mathcal{N}_1} \left\{ 1 - \frac{\bar{u}_{t-1,n}}{u_m} + \sqrt{\frac{2(1-\tilde{x}_t)\ln t}{k_{t-1,n}}} \right\}. \tag{3.20}$$

And the learning regret

$$R_T = \mathbb{E} \left[\sum_{t=1}^{T} x_t(u(t,n) - \mu^*) \right]$$

$$= u_m \mathbb{E} \left[\sum_{t=1}^{T} x_t \left\{ \left(1 - \frac{\mu^*}{u_m}\right) - \left(1 - \frac{u(t,n)}{u_m}\right) \right\} \right]. \tag{3.21}$$

Since $1 - \frac{\bar{u}_{t-1,n}}{u_m} \in [0,1]$, and $1 - \frac{u(t,n)}{u_m} \in [0,1]$, our task offloading problem is equivalent to the opportunistic bandit problem given in [11], equations (1–3), with equivalent definitions of utility function, decision making function and learning regret. By leveraging Lemma 7, Theorem 3 and Appendix C.2 in [11], we can get Theorem 3.1.

Theorem 3.1 shows that, under diverse input data size, the proposed ALTO algorithm can still effectively balance the tradeoff between exploration and exploitation, and achieve a bounded learning regret governed by $O(\ln T)$.

3.3.3 Performance Evaluation

In this subsection, we evaluate the learning regret and the average delay performance of the proposed ALTO algorithm through simulations. We start from a synthetic scenario to show the performance gain brought by occurrence-awareness and input-awareness respectively, and then use a realistic highway scenario for further verification.

3.3.3.1 Simulation Under Synthetic Scenario

We carry out simulations under a synthetic scenario in MATLAB. Consider a total number of $T = 1500$ time periods, with 1 TaV and 5 candidate SeVs which appear

Table 3.1 Candidate SeVs and their maximum CPU frequency

Index of SeV	1	2	3	4	5
F_n (GHz)	3.5	4.5	5.5	5	4
Epoch 1	✓	✓	–	–	–
Epoch 2	✓	✓	✓	✓	–
Epoch 3	✓	✓	✗	✓	✓

and disappear at different time. The distance between the TaV and each candidate SeV is within $[10, 200]$ m, and in each time period, the distance changes randomly within $[-10, 10]$ m. The wireless channel state is modeled by $h_{t,n}^{(u)} = h_{t,n}^{(d)} = A_0 l^{-2}$ according to [12], where l represents the distance between TaV and SeV and $A_0 = -17.8$ dB. Besides, the computation intensity of tasks is $\omega_0 = 1000$ cycles/bit, the transmit power $P = 0.1$ W, the channel bandwidth $W = 10$ MHz, and noise power $\sigma^2 = 10^{-13}$ W.

We first evaluate the impact of the occurrence time t_n on the learning regret and the average delay performance. The input data size of tasks is set as $x_0 = 0.6$ Mbits, and $\tilde{x}_t = 0$ for $\forall t$. There is a total number of 3 epochs, each having 500 time periods. The occurrence time and disappearance time of SeVs, together with their maximum CPU frequency F_n are shown in Table 3.1. In the second epoch, two SeVs with higher computing speed appear as candidates. In the third epoch, one SeV with suboptimal CPU frequency occurs, and SeV 3 with highest CPU frequency disappears. Also, the CPU frequency $f_{t,n}$ allocated to the TaV of interest is randomly selected from $20\% F_n$ to $50\% F_n$ in each time period.

As shown in Fig. 3.3a, we compare the learning regret of the proposed ALTO algorithm with the UCB algorithm proposed in [9], with padding function $\sqrt{\frac{\beta \ln t}{k_{t-1,n}}}$. In the first epoch, candidate SeVs appear from the very beginning, so that the proposed ALTO algorithm has the same padding function with UCB, and they perform the same. Starting from the second epoch, ALTO algorithm can effectively utilize the existing information of remaining SeVs, while learn the delay performance of the newly appeared SeVs faster. The occurrence awareness can reduce the learning regret by about 45%, compared with the classical UCB algorithm. As shown in Fig. 3.3b, the optimal delay is achieved by the genie-aided policy, regarded as the lower bound of the learning algorithm. We can see that the proposed ALTO algorithm converges faster to the optimal delay, and outperforms the UCB algorithm.

Then we evaluate the performance gain brought by the input awareness, and focus on a single epoch, by setting SeVs 1–4 as candidates for the whole 1500 time periods. The input data size x_t follows uniform distribution within $[0.2, 1]$ Mbits. As shown in Fig. 3.4, the proposed ALTO algorithm can reduce the learning regret by about 60% compared with the UCB algorithm, which verifies that considering the input data size of tasks can effectively balance the exploration and exploitation. Besides, the upper and lower thresholds should be carefully selected to further reduce the learning regret. Under our settings, threshold pairs $x^+ = 0.8, x^- = 0.4$ and $x^+ = x^- = 0.6$ perform slightly better than $x^+ = 0.9, x^- = 0.3$.

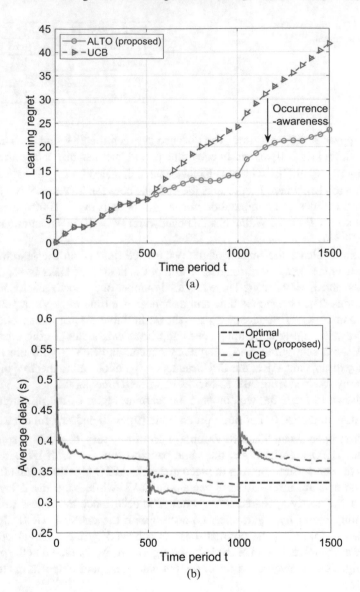

Fig. 3.3 Performance of ALTO under different SeV occurrence time and identical input data size. (**a**) Learning regret. (**b**) Average delay

3.3.3.2 Simulation Under Realistic Highway Scenario

In this part, we further carry out simulations under a realistic highway scenario. To emulate the real traffic flows, we use a traffic simulator called Simulation of Urban

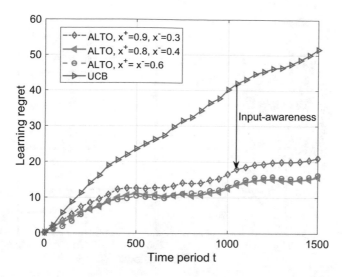

Fig. 3.4 Learning regret of ALTO under random input data size and identical candidate SeV set

Fig. 3.5 A 12 km stretch of
Beijing G6 Highway used for
simulations

Mobility (SUMO),[1] and download a 12 km stretch of Beijing G6 Highway from
Open Street Map (OSM).[2] The road is bi-directional four-lane, with two ramps in
between, as shown in the blue line in Fig. 3.5. We focus on a single TaV of interest,
while the arrival of SeVs each second from each ramp is modeled by Bernoulli
distribution, with the probability 0.1. The maximum allowed speed of vehicles is
set as 72 km/h.

We compare the proposed ALTO algorithm with UCB and two other benchmarks: (1) **Optimal Policy** is a genie-aided policy, where the TaV knows which
SeV performs the best in prior, and always selects the optimal one for each task. (2)
Random Policy is a naive policy where the TaV selects a SeV randomly in each
time period.

Figure 3.6 shows the average delay performance of ALTO, UCB and two
benchmarks. The proposed ALTO algorithm can achieve close-to-optimal average

---

[1]http://www.sumo.dlr.de/userdoc/SUMO.html

[2]http://www.openstreetmap.org/

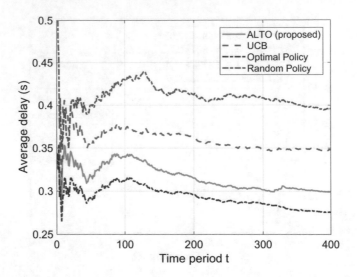

Fig. 3.6 Average delay performance of ALTO algorithm under a realistic highway scenario

delay, and outperforms the other two algorithms. To be specific, ALTO can improve the delay performance by about 15% compared with the UCB algorithm, through the joint consideration of occurrence time of vehicles and the input data size of tasks.

3.4 Reliability Enhancement Through Task Replication

In the VEC system, the dynamic network topology and wireless channels may make it unreliable to offload each task to a single TaV. Meanwhile, the computing resources are abundant due to the high density of SeVs. To further reduce the offloading delay and improve the service reliability of computation tasks, while exploiting the redundant computing resources of SeVs, we introduce *task repli-cation* method in this section: the replicas of each task are offloaded to multiple SeVs simultaneously, and processed by them without any cooperations. Once the computing result is transmitted back by one of these SeVs, the task is completed. The key idea of task replication is to exchange the redundant computing resources for quality of service (QoS) enhancement [13].

3.4.1 System Model and Problem Formulation

We still focus on the V-V offloading in the VEC system, and tasks are still offloaded in a distributed manner: each TaV makes decisions on the SeVs which should serve

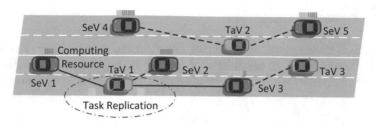

Fig. 3.7 An illustration of replicated task offloading in the VEC system

each task independently, without inter-vehicle coordinations. The task offloading procedure is similar to that illustrated in Sect. 3.2.1, while the major difference is that task replication is enabled. Figure 3.7 illustrates the task replication in the VEC system, where TaV 1 finds SeVs 1–3 as candidates, and offloads the current task replicas to SeV 1 and SeV 3.

3.4.1.1   Replicated Task Offloading Model

In each time period, the candidate SeV set of the TaV of interest is denoted by $\mathcal{N}_t$, and a subset of candidate SeVs $\mathcal{S}_t$ is selected by the TaV with $\mathcal{S}_t \subset \mathcal{N}_t$. Assume that the TaV selects a fixed number K of SeVs in each time period, i.e., $|\mathcal{S}_t| = K \leq |\mathcal{N}_t|$, and parameter K is related to the traffic conditions and task workloads, whose optimization is beyond our scope.

In time period t, the sum offloading delay $d_{\mathrm{sum}}(t, n)$ of each selected SeV $n \in \mathcal{S}_t$ is given by

$$d_{\mathrm{sum}}(t, n) = d_{\mathrm{up}}(t, n) + d_{\mathrm{com}}(t, n) + d_{\mathrm{dow}}(t, n), \qquad (3.22)$$

where $d_{\mathrm{up}}(t, n)$, $d_{\mathrm{com}}(t, n)$ and $d_{\mathrm{dow}}(t, n)$ are the uplink transmission delay, computation delay and downlink transmission delay, respectively, defined in Eqs. (3.2), (3.3), and (3.5).

The *actual offloading delay* of each task only depends on the fastest response among all selected SeVs:

$$\min_{n \in \mathcal{S}_t} d_{\mathrm{sum}}(t, n). \qquad (3.23)$$

However, we still require *all* the other selected SeVs to finish processing the task and provide the result feedbacks, to record the offloading delay for the learning purposes, which will be illustrated in detail in Algorithm 3.2.

3.4.1.2 Problem Formulation

Given a total number of T time periods, our objective is to minimize the average actual offloading delay of tasks, by deciding the subset of candidate SeVs which should be selected by the TaV to serve each task. The problem is formulated as:

$$\textbf{P4:} \quad \min_{\mathscr{S}_1,\ldots,\mathscr{S}_T} \frac{1}{T} \sum_{t=1}^{T} \min_{n \in \mathscr{S}_t} d_{\text{sum}}(t, n). \tag{3.24}$$

As discussed in Sect. 3.2.2, we still assume that some kinds of state information, including the CPU frequency $f_{t,n}$, transmission rates $r_{t,n}^{(u)}$, $r_{t,n}^{(d)}$ and interference $I_{t,n}^{(u)}$ are unknown to the TaV in prior, and the TaV learns the delay performance of candidate SeVs while offloading tasks, which is called **learning while offloading**. Specifically, up till time period t, the TaV gets delay records $\{d_{\text{sum}}(1, n), n \in \mathscr{S}_1\}$, $\ldots$, $\{d_{\text{sum}}(t - 1, n), n \in \mathscr{S}_{t-1}\}$, estimates the delay performance at the current time period, and selects a subset $\mathscr{S}_t = \arg\min_{\mathscr{S} \in \mathscr{N}_t} \min_{n \in \mathscr{S}} d_{\text{sum}}(t, n)$ to offload the task replicas.

The problem formulated in (3.24) has a non-linear form of objective function $\min_{n \in \mathscr{S}_t} d_{\text{sum}}(t, n)$, which makes the learning process more difficult than that without task replication. To further simplify the problem, we assume that tasks are identical across time, with equal input, output data size $x_t = x_0$, $y_t = y_0$ and computation intensity $w_t = w_0$ for $\forall t$. Then the sum offloading delay to SeV n in time period t can be written as

$$d_{\text{sum}}(t, n) = \frac{x_0}{r_{t,n}^{(u)}} + \frac{y_0}{r_{t,n}^{(d)}} + \frac{w_0}{f_{t,n}}, \tag{3.25}$$

and thus the sum offloading delay can directly reflect the comprehensive communication and computing service capability of each candidate SeVs, which the TaV needs to learn.

We aim at designing a learning algorithm that minimizes the expectation of the average actual offloading delay, given by

$$\textbf{P5:} \quad \min_{\mathscr{S}_1,\ldots,\mathscr{S}_T} \frac{1}{T} \mathbb{E}\left[\sum_{t=1}^{T} \min_{n \in \mathscr{S}_t} d_{\text{sum}}(t, n) \right]. \tag{3.26}$$

3.4.2 Learning-Based Task Replication Algorithm

As shown in Algorithm 3.2, we design a learning-based task replication algorithm (LTRA), which guides the TaV to learn the delay performance of candidate SeVs while offloading tasks. The proposed algorithm is based on the combinatorial MAB (CMAB) theory [14]. Compared with classical MAB problem, the major

---

Algorithm 3.2 Learning-based task replication algorithm

1: **for** $t = 1, \ldots, T$ **do**
2:     **if** Any SeV $n \in \mathcal{N}_t$ has not connected to the TaV **then**
3:         Connect to any subset $\mathcal{S}_t \in \mathcal{N}_t$ once, with $n \in \mathcal{S}_t$.
4:         Update empirical CDF $\hat{F}_n$ of normalized delay $\tilde{d}(t, n)$ and selected times $k_{t,n}$ for each
        $n \in \mathcal{S}_t$.
5:     **else**
6:         For each $n \in \mathcal{N}_t$, define CDF $\underline{G}_n$ as

$$\underline{G}_n(x) = \begin{cases} 0 & x = 0, \\ \min\left\{\hat{F}_n(x) + \sqrt{\frac{\beta \ln(t - t_n)}{k_{t-1,n}}}, 1\right\} & 0 < x \leq 1 \end{cases}. \tag{3.27}$$

7:         Select a subset of K candidate SeVs, such that

$$\mathcal{S}_t = \arg\min_{\mathcal{S} \subset \mathcal{N}_t} \mathbb{E}_{\underline{D}}\left[\min_{n \in \mathcal{S}} d_{\text{sum}}(t, n)\right], \tag{3.28}$$

        where $\underline{D}_n$ is the distribution of CDF $\underline{G}_n$, and $\underline{D} = \underline{D}_1 \times \underline{D}_2 \times \ldots \times \underline{D}_{|\mathcal{N}_t|}$.
8:         Offload the task replica to all the SeV $\forall n \in \mathcal{S}_t$.
9:         Observe delay $d_{\text{sum}}(t, n)$ for each $n \in \mathcal{S}_t$.
10:       Update $\hat{F}_n$ and $k_{t,n} \leftarrow k_{t-1,n} + 1$ for each $n \in \mathcal{S}_t$.
11:     **end if**
12: **end for**

---

difference of the CMAB problem is that the player can try a *subset* of actions at each time, and observe the losses of all the selected actions. The loss of the system is a function of the losses of selected actions, and the objective is to minimize the cumulative loss over time. The CMAB problem still needs to balance the tradeoff between *exploration and exploitation* during the learning process: to explore different combinations of actions and acquire more information of their loss distributions, or to exploit the existing information and select the subset of actions with empirically lowest loss.

Our problem defined in (3.26) is similar to the CMAB problem: each candidate SeV can be regarded as an action with unknown delay distribution, while the TaV corresponds to the player who chooses a subset $\mathcal{S}_t$ of SeVs in each time period t. The TaV observes the delay performance of all the selected SeVs, and the loss of the system is the actual offloading delay $\min_{n \in \mathcal{S}_t} d_{\text{sum}}(t, n)$, which is a *non-linear* function of the observed delay set.

A major feature of our task replication problem is that, the candidate SeV set $\mathcal{N}_t$ changes over time due to the movements of vehicles, and the TaV cannot know in prior when SeVs may appear and disappear within its communication range, and how long they may act as candidates. How to learn the delay performance of the dynamic candidate SeVs efficiently is a key challenge.

We thus modify the existing CMAB algorithm in [15], by taking the time varying feature of candidate SeV set into consideration, and propose the LTRA, as shown in Algorithm 3.2. We normalize the observed delay of each candidate SeV $n \in \mathcal{S}_t$ as

$\tilde{d}(t, n) = \frac{d_{\text{sum}}(t,n)}{d_{max}}$, with $\tilde{d}(t, n) \in [0, 1]$, where d_{max} is the maximum delay allowed to offload each task. If the computation result is not successfully transmitted back from SeV n within time duration d_{max}, we consider that the task is failed due to the insufficient computing resource or transmission failure, and set the observed delay $d_{\text{sum}}(t, n) = d_{max}$ for learning purpose. Let the empirical probability distribution function (PDF) of the normalized delay $\tilde{d}(t, n)$ of SeV n be $\hat{D}_n$, and the cumulative distribution function (CDF) of $\hat{D}_n$ be $\hat{F}_n$. The number of tasks offloaded to SeV n up till time period t is denoted by $k_{t,n}$, and the time that SeV n occurs as a candidate is denoted by t_n. And let β be a constant factor.

In Algorithm 3.2, Lines 2–4 are the initialization phase, which happens not only at the beginning of the learning process, but whenever any new SeV appears. The TaV chooses a subset of K SeVs which includes at least one newly occurred SeV, and gets an initial estimation of its delay performance.

Lines 6–10 are the main loop of the learning algorithm. Different from the ALTO algorithm where the TaV learns the mean delay performance of each candidate SeV, the task replication problem brings a non-linear feature to the objective function $\min_{n \in \mathscr{S}_t} d_{\text{sum}}(t, n)$. Therefore, the offloading decision $\mathscr{S}_t$ depends on the *joint delay distribution* of candidate SeVs, rather than their individual mean values. The proposed LTRA algorithm learns the empirical CDF $\hat{F}_n$ and PDF $\hat{D}_n$ of candidate SeVs, and makes offloading decisions according to them.

In Line 6, we define the utility function $\underline{G}_n(x)$ in (3.27), which is a numerical upper confidence bound of realistic delay CDF of each candidate SeV n. The utility function can balance the tradeoff between exploration and exploitation in the learning process: when the number of selected times $k_{t,n}$ is small, the TaV tends to *explore* different SeVs to learn their good delay estimations. On the other hand, when $k_{t,n}$ is large, the latter term $\sqrt{\frac{\beta \ln(t - t_n)}{k_{t-1,n}}}$ is small, such that the TaV tends to *exploit* the SeVs with better empirical delay performance. The occurrence time t_n is also considered in the padding term, leading the TaV to explore newly appeared SeVs while exploiting the empirical information of existing SeVs.

In Line 7, the subset of candidate SeVs is selected according to (3.28), which is a minimum element problem that minimizes the expected actual offloading delay [16]. Here, $\underline{D}_n$ is the PDF of CDF $\underline{G}_n$, and $\underline{D} = \underline{D}_1 \times \underline{D}_2 \times \ldots \times \underline{D}_{|\mathcal{N}_t|}$ is the joint distribution of all candidate SeVs. Then in Lines 8–10, the TaV offloads the replicas of the task to the selected SeVs $\forall n \in \mathscr{S}_t$, observes the offloading delay through result feedbacks, and updates the empirical CDF $\hat{F}_n$ and the number of offloaded tasks $k_{t,n}$.

3.4.2.1   Implementation Considerations

Due to the continuous value of the offloading delay, the proposed LTRA algorithm may lead to high storage usage and computational complexity when t grows large. Specifically, the TaV may observe a different value of $d_{\text{sum}}(t, n)$ in each time period, so that the storage required to record the CDF $\hat{F}_n$ of each SeV n is $O(t)$, and the minimum element problem in Line 7 is of higher computational complexity to be solved. Therefore, it may not be easy to implement the proposed LTRA algorithm directly in the realistic VEC system.

One solution is to discretize the CDF $\hat{F}_n$ to be $\tilde{F}_n$, by partitioning the normalized delay $\tilde{d}(t, n) \in [0, 1]$ into l segments with equal interval $\frac{1}{l}$. The support of $\tilde{F}_n$ is $\{\frac{1}{l}, \frac{2}{l}, \ldots, 1\}$. If the normalized delay $\tilde{d}(t, n)$ belongs to $\left(\frac{i-1}{l}, \frac{i}{l}\right]$, the TaV updates $\tilde{F}_n$ regarding $\tilde{d}(t, n) = \frac{i}{l}$.

3.4.2.2   Performance Guarantee

We still adopt the *learning regret* to characterize the performance of the proposed LTRA algorithm. To simplify the performance analysis, we only focus on a single epoch with totally T time periods, during which the candidate SeV set $\mathcal{N}_t$ remains constant, denoted by $\mathcal{N}$. We will verify through simulations later that without this assumption, the proposed LTRA algorithm can still perform well.

Define the delay vector $\boldsymbol{d}_t = (d_{\text{sum}}(t, 1), \ldots, d_{\text{sum}}(t, N))$, where $N = |\mathcal{N}|$ is the number of candidate SeVs. The actual offloading delay can be represented by $L(\boldsymbol{d}_t, \mathcal{S}_t) = \min_{n \in \mathcal{S}_t} d_{\text{sum}}(t, n)$. Define $\mu_{\mathcal{S}} = \mathbb{E}_t[L(\boldsymbol{d}, \mathcal{S})]$ as the expected actual offloading delay of selecting a subset $\mathcal{S}$ of candidate SeVs. Let $\mathcal{S}^* = \arg\min_{\mathcal{S} \subset \mathcal{N}} \mu_{\mathcal{S}}$ be the optimal subset of SeVs, and $\mu_{\mathcal{S}^*} = \min_{\mathcal{S} \subset \mathcal{N}} \mu_{\mathcal{S}}$.

The learning regret is the expected cumulative loss of actual offloading delay caused by learning, compared with the genie-aided optimal policy, defined as

$$R_T = \mathbb{E}\left[\sum_{t=1}^{T} L(\boldsymbol{d}_t, \mathcal{S}_t)\right] - T\mu_{\mathcal{S}^*}, \qquad (3.29)$$

For any suboptimal SeV subset $\mathcal{S} \in \mathcal{N}$, define the expected difference of actual delay as

$$\Delta_{\mathcal{S}} = (\mu_{\mathcal{S}} - \mu_{\mathcal{S}^*})/d_{max}. \qquad (3.30)$$

Let $\mathcal{N}_s$ be the set of SeVs which are contained in at least one suboptimal SeV subset, and

$$\Delta_n = \min\{\Delta_{\mathcal{S}} | \mathcal{S} \subset \mathcal{N}, n \in \mathcal{S}, \mu_{\mathcal{S}} > \mu_{\mathcal{S}^*}\}, \qquad (3.31)$$

Then the learning regret of the proposed LTRA algorithm can be upper bounded as follows.

Theorem 3.2 *Let* $\beta = \frac{2}{3}$, *then* R_T *is upper bounded by.*

$$R_T \leq d_{max}\left(C_1 K \sum_{n \in \mathcal{N}_s} \frac{\ln T}{\Delta_n} + C_2\right), \qquad (3.32)$$

where $C_1 = 2136$ and $C_2 = \left(\frac{\pi^2}{3} + 1\right) N$ are two constants.

Proof Our task replication problem can be transformed as the standard CMAB problem investigated in [15].

First, the objective functions are equivalent, since

$$\min_{\mathscr{S}_1,\ldots,\mathscr{S}_T} \frac{1}{T} \sum_{t=1}^{T} \min_{n \in \mathscr{S}_t} d(t, n)$$

$$= d_{max} \min_{\mathscr{S}_1,\ldots,\mathscr{S}_T} \frac{1}{T} \sum_{t=1}^{T} \min_{n \in \mathscr{S}_t} \tilde{d}(t, n)$$

$$\Leftrightarrow \max_{\mathscr{S}_1,\ldots,\mathscr{S}_T} \frac{1}{T} \sum_{t=1}^{T} \left[\max_{n \in \mathscr{S}_t} \left(1 - \tilde{d}(t, n) \right) \right]. \tag{3.33}$$

Since $\tilde{d}(t, n) \in [0, 1]$, the reward function $R(\boldsymbol{d}_t, \mathscr{S}_t) = \max_{n \in \mathscr{S}_t} \left(1 - \tilde{d}(t, n)\right) \in [0, 1]$ and it is monotone, satisfying assumption 2 and 3 [15] with upper bound $M = 1$.

Second, the utility function $\underline{G}_n(x)$ defined in (3.27) can be transformed to CDF $\underline{F}_n(x)$ defined in the SDCB algorithm in [15]. Define $\hat{F}_n(x)$ as the CDF of $\tilde{d}(t, n)$, and $\hat{F}_n'(x)$ the CDF of $1 - \tilde{d}(t, n)$. It is easy to see that $\hat{F}(x) = 1 - \hat{F}'(1 - x)$. Thus

$$\underline{G}_n(x) = 1 - \underline{F}_n(1 - x)$$

$$= 1 - \begin{cases} \max\left\{ \hat{F}_n'(1 - x) - \sqrt{\frac{\beta \ln t}{k_{t-1,n}}}, 0 \right\} & 0 \leq 1 - x < 1, \\ 1 & 1 - x = 1 \end{cases}$$

$$= 1 - \begin{cases} \max\left\{ 1 - \hat{F}_n(x) - \sqrt{\frac{\beta \ln t}{k_{t-1,n}}}, 0 \right\} & 0 < x \leq 1, \\ 1 & x = 0 \end{cases}$$

$$= \begin{cases} 0 & x = 0, \\ \min\left\{ \hat{F}_n(x) + \sqrt{\frac{\beta \ln t}{k_{t-1,n}}}, 1 \right\} & 0 < x \leq 1 \end{cases}. \tag{3.34}$$

Substitute the reward upper bound $M = 1$, and let $\alpha = 1$ in Theorem 1 in [15], Theorem 3.2 can be derived.

3.4.3 Performance Evaluation

We use the same Beijing G6 Highway scenario as in Sect. 3.3.3.2, to evaluate the proposed LTRA algorithm. Compared with the previous settings, the major differences in this simulation include:

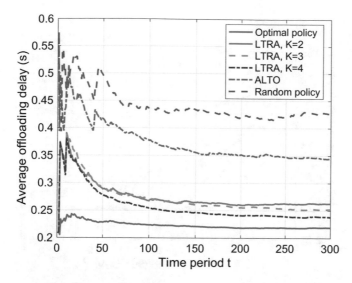

Fig. 3.8 Average offloading delay of LTRA

(1) The input data size of tasks is fixed as $x_0 = 1$ Mbits.

(2) The maximum CPU frequency F_n of each SeV is uniformly selected within [2, 8] GHz, and the CPU frequency allocated to each TaV is randomly distributed within $[0, 60\%]F_n$. Under these settings, the VEC environment is more dynamic than that in the previous simulations.

Other default parameters include: the number of discretization segments l is set as 50, and $\beta = 0.6$.

We compare the average delay of LTRA with $K = 2, 3, 4$ task replicas, with the ALTO algorithm (without task replication), Random Policy and Optimal Policy (introduced in Sect. 3.3.3.2). The density ratio of TaV and SeV is 0.25. As shown in Fig. 3.8, when the TaV faces a more dynamic task offloading environment, task replication can effectively improve the average delay performance compared with the ALTO algorithm. While ALTO algorithm can only achieve the average delay 0.35 s, LTRA can improve the delay to about 0.25 s. And the average delay is much closer to the optimal genie-aided policy with larger number of task replicas.

In Fig. 3.9, there are 2 TaVs surrounded by 2 to 10 SeVs, and the average delay and service reliability are evaluated. As shown in Fig. 3.9a, when the number of candidate SeVs increases, the average delay of LTRA decreases, since task replication enables the TaV to exploit the redundant computing resources. However, the ALTO algorithm cannot make use of the sufficient computing resources to reduce offloading delay. Another observation is that, more task replicas cannot always improve the delay performance. When there are less than 4 candidate SeVs, LTRA with $K = 4$ suffers higher average delay than that with $K = 2$, due to the inefficient use of limited computing resources. The completion ratio of tasks given deadline 0.6 s is shown in Fig. 3.9b. When there are sufficient computing resources,

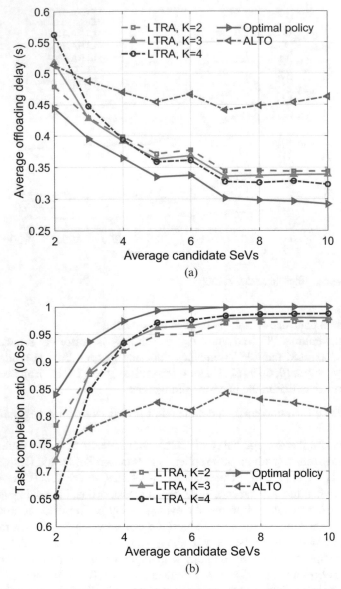

Fig. 3.9 Performance of LTRA under different number of candidate SeVs. (**a**) Average offloading delay. (**b**) Task completion ratio

the service reliability is improved substantially by task replication, reaching over 98%, while ALTO algorithm can only achieve around 80%.

We then evaluate the delay performance of LTRA under different TaV and SeV density ratio. As shown in Fig. 3.10, the increasing TaV and SeV density ratio means that more TaVs are competing for the computing resources. When the density ratio

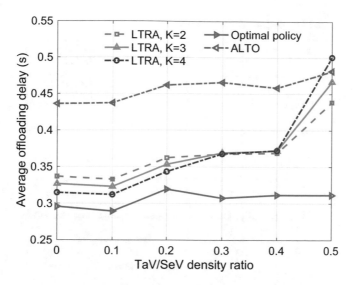

Fig. 3.10   Average offloading delay of LTRA under different TaV and SeV density ratio

is below 0.3, LTRA with $K = 4$ performs the best. As the density ratio grows higher than 0.4, LTRA with fewer number of task replicas reach lower delay performance. Therefore, the number of task replicas should be optimized based on the traffic conditions.

Finally, we evaluate how the discretization level l affects the average delay and the runtime of the algorithm. As shown in Fig. 3.11, when the discretization level l is low, the average delay performance of LTRA is poor and fluctuates drastically, since the discretized delay CDF loses a lot of information. When the discretization level is too high, the runtime of LTRA is long, but the delay performance does not improve much. To obtain good estimates of the realistic delay distributions and reduce the runtime, we should carefully select the discretization level. For example, under our simulation settings, l should be about 40.

3.5   System Level Simulation Platform

To further emulate the vehicular environment and test our proposed algorithms, we build a system level simulation platform based on an open source Veins[3] (Vehicles in Network Simulations).

As shown in Fig. 3.12, Veins enables large-scale vehicular simulations, which integrates a network simulator OMNeT++[4] and a traffic simulator SUMO, and supports vehicular communication protocols such as DSRC protocol. It encapsulates

---

[3] http://veins.car2x.org/

[4] https://www.omnetpp.org/documentation

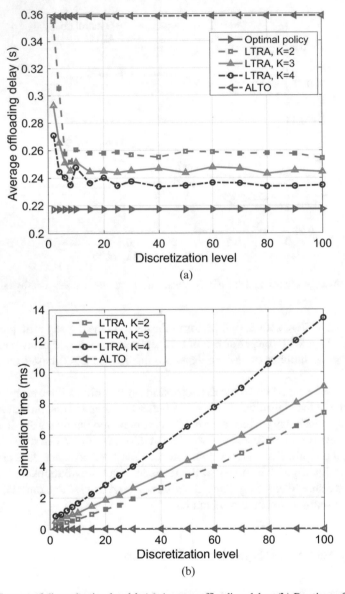

Fig. 3.11 Impact of discretization level l. (**a**) Average offloading delay. (**b**) Runtime of LTRA

the connection to SUMO through Traffic Control Interface (TraCI) in a mobility module, and contains realistic channel models and obstacle shadowing.

To simulate the VEC system, we define TaV nodes and SeV nodes in the simulation platform, and the task offloading procedures. Veins is an event-based network simulator, in which the actions of nodes are triggered by different events such as task generation and data reception. The main event messages of TaVs and SeVs and their corresponding handlers are summarized in Table 3.2, and the flow

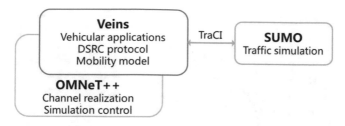

Fig. 3.12 Architecture of the simulation platform

Table 3.2 Event messages and handlers in the simulation platform

Node type	Message	Handler
SeV	Send beacons every 1 s	sendBeacon()
	Receive task information and process it	processBrief()
	Receive input data and calculate computing delay	processTask()
	Transmit back output data	sendResult()
TaV	Process beacons and select candidate SeVs	handleBeacon()
	Generate, schedule and offload tasks	handleOffload()
	Receive result feedback and update delay	updateResult()

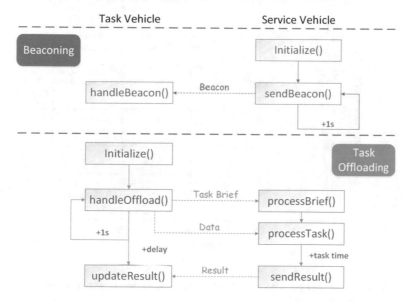

Fig. 3.13 Flow chart of the simulation platform

chart of simulation is shown in Fig. 3.13. Each SeV broadcasts a beacon message
through sendBeacon() every 1 s, reporting its identity, location, moving direction
and speed. TaVs receive beacon messages from surrounding SeVs and select candidate SeVs, realized by handleBeacon(). The handleOffload() function generates
tasks and runs the task offloading algorithms such as ALTO and LTRA to offload

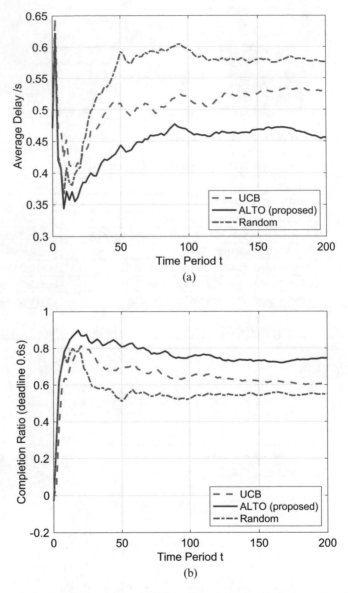

Fig. 3.14 Performance of the proposed ALTO algorithm in system level simulation platform. (**a**) Average offloading delay. (**b**) Task completion ratio

tasks. The processTask() and processBrief() functions handle the data reception and data processing at SeVs, respectively. And finally results are transmitted back from SeVs and received by TaVs through sendResult() and updateResult() functions.

Figures 3.14 and 3.15 show the average offloading delay and task completion ratio of the proposed ALTO, LTRA algorithms. We can see that in the realistic

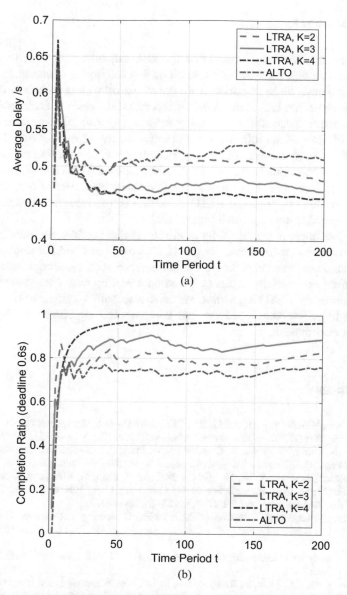

Fig. 3.15 Comparison of the proposed ALTO and LTRA algorithms in system level simulation platform. (**a**) Average offloading delay. (**b**) Task completion ratio

vehicular environment, the proposed ALTO algorithm still outperforms the exiting learning-based UCB algorithm and naive random policy. And task replication can further improve the delay performance and service reliability.

3.6 Summary

In this chapter, we have investigated the task offloading problem in the VEC system, and proposed a learning while offloading solution to minimize the average offloading delay. We have started from the single offloading case, and developed an ALTO algorithm based on the MAB theory, which enables each TaV to learn the delay performance of its surrounding SeVs in a distributed manner, to deal with the lack of global state information. In particular, we have addressed the critical challenges of vehicular task offloading, including the time varying candidate SeV set and the task workloads, and proved that the proposed ALTO algorithm has strong performance guarantee. Extensive simulations have been carried out, and results have shown that the proposed ALTO algorithm can reduce the average delay by 15%, compared to existing MAB algorithms.

To further improve the reliability of the computing services, we have introduced the task replication method, and designed LTRA based on CMAB. Simulations have shown that when there are sufficient computing resources, the service reliability can be improved substantially by task replication, reaching over 98%, compared with 80% achieved by ALTO algorithm. We have also built a system level simulation platform based on Veins, and verified the proposed algorithms in the realistic vehicular environment.

References

1. J. S. Choo, M. Kim, S. Pack, and G. Dan, "The software-defined vehicular cloud: A new level of sharing the road," *IEEE Veh. Technol. Mag.*, vol. 12, no. 2, pp. 78–88, Jun. 2017.
2. N. Lu, N. Cheng, N. Zhang, X. Shen, and J. W. Mark, "Connected Vehicles: Solutions and Challenges," *IEEE Internet Things J.*, vol. 1, no. 4, pp. 289–299, Aug. 2014.
3. M. Sookhak, F. R. Yu, Y. He, H. Talebian, N. Safa, N. Zhao, M. Khan, and N. Kumar, "Fog Vehicular Computing: Augmentation of Fog Computing Using Vehicular Cloud Computing," *IEEE Veh. Technol. Mag.*, vol. 12, no. 3, pp. 55–64, Sept. 2017.
4. C. Huang, R. Lu, and K. K. R. Choo, "Vehicular Fog Computing: Architecture, Use Case, and Security and Forensic Challenges," *IEEE Commun. Mag.*, vol. 55, no. 11, pp. 105–111, Nov. 2017.
5. 3GPP, "Study on enhancement of 3GPP support for 5G V2X services," 3GPP TR 22.886, V15.1.0, Mar. 2017.
6. Y. Sun, X. Guo, S. Zhou, Z. Jiang, X. Liu, and Z. Niu, "Learning-based task offloading for vehicular cloud computing systems," in *Proc. IEEE Int. Conf. Commun. (ICC)*, Kansas City, MO, USA, May 2018.
7. J. B. Kenney, "Dedicated short-range communications (DSRC) standards in the United States," *Proceedings of the IEEE*, vol. 99, no. 7, pp. 1162–1182, Jul. 2011.
8. Y. Mao, C. You, J. Zhang, K. Huang, and K. B. Letaief, "A survey on mobile edge computing: The communication perspective," *IEEE Commun. Surveys Tuts.*, vol. 19, no. 4, pp. 2322–2358, 2017.
9. P. Auer, N. Cesa-Bianchi, and P. Fischer, "Finite-time analysis of the multiarmed bandit problem," *Machine learning*, vol. 47, no. 2–3, pp. 235–256, 2002.

10. Z. Bnaya, R. Puzis, R. Stern, and A. Felner, "Social network search as a volatile multi-armed bandit problem," *HUMAN*, vol. 2, no. 2, pp. 84–98,

11. H. Wu, X. Guo, and X. Liu, "Adaptive exploration-exploitation tradeoff for opportunistic bandits." *International Conference on Machine Learning (ICML)*, Stockholm, Sweden, Jul. 2018.

12. M. Abdulla, E. Steinmetz, and H. Wymeersch, "Vehicle-to-vehicle communications with urban intersection path loss models," in *Proc. IEEE Global Commun. Conf. (GLOBECOM)*, Washington, DC, USA, Dec. 2016.

13. Y. Sun, J. Song, S. Zhou, X. Guo, and Z. Niu, "Task replication for vehicular edge computing: A combinatorial multi-armed bandit based approach," *IEEE Global Commun. Conf. (GLOBECOM)*, Abu Dhabi, UAE, Dec. 2018.

14. W. Chen, Y. Wang, and Y. Yuan. "Combinatorial multi-armed bandit: General framework and applications," *Int. Conf. on Machine Learning (ICML)*, Atlanta, GA, USA, Jun. 2013.

15. W. Chen, W. Hu, F. Li, J. Li, Y. Liu, and P. Lu, "Combinatorial multi-armed bandit with general reward functions," *Advances in Neural Information Processing Systems*, vol. 29, 2016.

16. A. Goel, S. Guha, and K. Munagala, "How to probe for an extreme value," *ACM Trans. on Algorithms*, vol. 7, no. 1, Nov. 2010.

Chapter 4
Intelligent Network Access System for Vehicular Real-Time Service Provisioning

With mobile operating systems becoming increasingly common in vehicles, it is undoubted that vehicular demands for real-time Internet access would get a surge in the soon future. The vehicular ad hoc network (VANET) offloading represents a promising solution to the overwhelming traffic problem engrossed to cellular networks. With a vehicular heterogeneous network formed by a cellular network and VANET, efficient network selection is crucial to ensuring vehicles' quality of service (QoS), avoiding network congestions and other performance degradation. To address this issue, we develop an intelligent network access system using the control theory to provide seamless vehicular communication. Specifically, our system comprises two components. The first component recommends vehicles an appropriate network to access by employing an analytic framework which takes traffic status, user preferences, service applications and network conditions into account. In the second one, a distributed automatic access engine is developed by utilizing a learning method, which enables individual vehicles to make access decisions based on access recommender, local observation and historic information. Lastly, simulations show that our proposal can effectively select the optimum network to ensure the QoS of vehicles, and network resource is fully utilized without network congestions in the meantime.

4.1 Network Selection for Heterogeneous Vehicular Networks

With the first Volvo car equipped CarPlay (previously known as iOS in cars) shown in April 2014 on New York Auto Show, Google has also announced that the Android Auto vehicles are expected to begin rolling off lots by the end of 2014 [1]. Therefore, it is foreseeable that Internet access, e.g., social networking, real-time traffic report and navigation, would become a standard feature of future motor vehicles. As the

© Springer Nature Switzerland AG 2019
L. Xiao et al., *Learning-based VANET Communication and Security Techniques*, Wireless Networks, https://doi.org/10.1007/978-3-030-01731-6_4

cellular infrastructure still represents the dominant access methods for ubiquitous connections, the main way to provide reliable and ubiquitous Internet access to vehicles will still be through the cellular-based access technologies [2, 3], such as Long Term Evolution (LTE) and emerging 5G cellular network. However, as vehicular mobile data traffic is undergoing an explosive growth, simply using the cellular infrastructure for vehicle Internet access may result in an increasingly severe traffic overloading issue, which eventually would degrade the service performance of both traditional smartphone and vehicular mobile users.

4.1.1 Heterogeneous Vehicular Networks

To offload the cellular traffic, the city-wide WLANs have been developed in many countries, which are possible to be extended to serve vehicular users using the emerging IEEE802.11p standard [4], known as WAVE (Wireless Access in a Vehicular Environment). The WAVE protocols are designed in the 5.850–5.925 GHz band, the Dedicated Short Range Communications (DSRC) spectrum band, for VANETs which can provide in-vehicle users with a variety of applications for safety, traffic efficiency and infotainment [5]. As a result, the wide deployment of different wireless technologies, in particular 3G and DSRC, in combination, and the advanced vehicles with multiple network interfaces equipped, would allow in-vehicle users to access to different real-time services at anywhere anytime from any networks [6, 7].

4.1.2 Network Selection Problem

Network selection is important for handover in heterogeneous wireless environment. A perfect network selection/handover mechanism should ensure the applications running in mobile devices are uninterrupted. To provide continuous end-to-end services, the mechanism should also decrease the packet loss rate and the transmission delay through selecting a channel with better quality [8]. Generally, the network selection mechanism consists of three steps [9]: information gathering, network selection and access execution. In the process of network selection, the main information can be grouped into four categories: (1) the information about networks, such as the quality of channel, delay, cost and so on; (2) the information about nodes, such as the location, velocity and so on; (3) the information about users, such as the requirements, preferences and so on; (4) the information about services, such as the specific application, the quality of service and so on. Based on these information, authors classified vertical handover decision strategies into five main categories: (a) function based, (b) user centric, (c) multiple-attribute decision, (d) fuzzy logic and neural networks based and (e) context aware strategies [10–14].

The function based decision can integrate some specific parameters of networks to form an aggregate multi-attributes utility function. The weighted utility based mathematical modeling for network selection is concluded in [15], which is simple and fast, but the accuracy of its decision making is not high. User centric mechanism always regards the requirement of cost or QoS as the goal to evaluate [16] the mechanism's performance, which has the advantage of simple realization, but its single target feature results in the limit application for user satisfaction related and some non-real-time services. The network selection mechanism handles selection problem among many types of networks, which forms a multiple attribute decision making algorithm (MADM) [17]. MADM performs on the network side to assist the terminal to select the top candidate network from the set of alternatives (usually with contrary natures), which needs a high accuracy of the information. Typical resolutions of the MADM problem consists Analytic Hierarchy Process (AHP) and Grey Relational Analysis (GRA). Kassar et al. [10] presents an AHP based network selection algorithm to make a decision between UMTS and WLAN. In the algorithm, users' preferences and service requirements for QoS are valued based on their contributions to the final goal through AHP. Then, GRA is integrated to rank the alternatives, and the highest ranked network is selected as the decision result. Using the fuzzy logic and neural network, the network selection algorithm based on multi-criteria and multi-target is proposed, which can provide multiple real-time and non-real-time services [18]. It can solve the inefficiency of decision because of the inaccurate data. Additionally, based on the toleration of the errors of attribute information, fuzzy network can merge the information of each attribute to make the decision with a modest accuracy. In order to make an intelligent decision, context aware vertical handoff algorithms is proposed, which is based on a context repository and an adaptability manager [19]. Except that the context repository may be error, this method needs frequency transmissions between terminals and networks, which may result in additional overhead on the communication link. Apart from these five aspects, network selection issue can also be converted into a game problem between the users and networks, and an evolutionary game approach is proposed for dynamic network selection in [20]. However, the game theory method has high computational complexity and it is not suitable for communication networks with strict requirements on latency. In [21], Hasib also describes a mobility adaptive network selection scheme, in the context of WWAN and WLAN, based on the theory of Markov chain. All of the above methods are used in traditional heterogeneous network selection/handover issues, such as the networks with 3G, WLAN, and WiMax.

4.1.3 Network Selection in Vehicular Networks

To provide in-vehicle users with desirable services, the essential engineering challenge is how to ensure the desirable QoS of vehicular users during the roaming as different access technologies have diverse available bandwidth, delay and communication cost. In addition, facing the dynamic connections of vehicular users, how to fully utilize the network resources without violating the QoS requirements of in-vehicle users also represents fundamental challenges. To summarize, an efficient network selection scheme, which guides vehicles to always connect to the best network with guaranteed QoS whereas fully utilize the network resource, is urgently desirable.

From above section, we knows that many network selection algorithms have been studied. However, considering network selection mechanisms based on vehicular communications have not many researches, we reference other heterogeneous wireless network selection algorithms. Kousalya et al. proposes a handover algorithm based on prediction mechanism [22]. An joint optimized algorithm is proposed in [23], which aims at maximizing vehicle nodes' battery life and minimizing vehicles' cost meantime. However, the energy consumption always not be the most important factor in the process of network selection decision making. In [24], AHP is used to solve the vertical handover problem with multiple constraint environment for ITS, and user preferences and network contexts are taken into consideration. However, researches for network selection in urban traffic environment are really few. In the context of future wireless networks, our research aims to design an efficient network access system for vehicular users' real-time services provisioning. Consider the hierarchical architecture shown in Fig. 1.1, and the system model and some preliminaries is given below.

4.1.3.1   Traffic Distribution Model

At first, we observe the urban traffic distribution after processing the GPS data of taxies in Shanghai, China. Figure 4.1 is the processing result of an area which takes

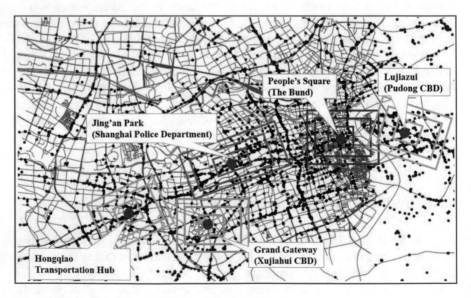

Fig. 4.1 Result of the GPS data of taxies in Shanghai

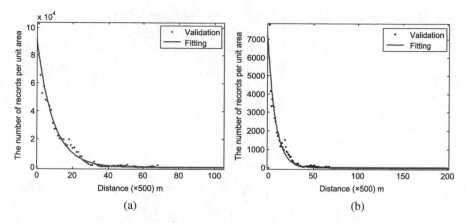

Fig. 4.2   The fitting curve of data. (**a**) Data of the whole day. (**b**) Data of two hours

[121.347, 31.1505], [121.540, 31.2918] as two end points of a diagonal and in the internal of 1 min through map matching. In Fig. 4.1, blue lines in the bottom layer represent the road structure of Shanghai, and each red point represents a data of a vehicle. We can observe that the distribution of vehicles has differences with different region. That is to say, if the region with much number of data is taken as a social spot (SP), the number of data around the SP is gradually decreased. As the five big red points shown in the figure, they are SPs of Shanghai, and the number of data is gradually decreased from the center of SPs to their boundary.

After further analysis, we find that the spatial stationary distribution of vehicles follows a power-law decay. As shown in Fig. 4.2, we take People Square as the center, the horizontal ordinate represents the distance from the center and the vertical ordinate represents the number of data in a unit area. We can observe that the number of data from the SP to their boundary is well fitted by an exponential curve. Figure 4.2a is the result of the data in the whole day which is fitted to $91640e^{-0.196}$, and Fig. 4.2b is the result of the data in two hours which is fitted to $6442e^{-0.1175}$.

Hence, the characteristics of urban traffic [25] can be summed up as follows.

- The mobility trace of vehicle has social features. Simply, the mobility region of vehicle is restricted and associated with SPs. Notably, vehicles only move within a bounded region related to the social life of the driver mostly.
- The urban traffic distribution expresses difference in different regions. The density of vehicles within the proximity area of SPs is much higher than the average and follows the power-low decay distribution.

Based on these features, we model the distribution of vehicles in the urban traffic environment through establishing *Tier* model in the follow-up work.

Fig. 4.3 The grid-like street pattern

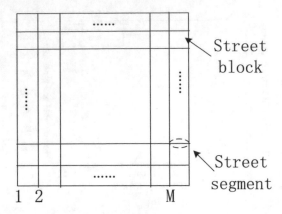

4.1.3.2 **Problem Definition**

Normally, the urban area is covered by heterogeneous wireless networks including cellular network and VANET. Different characteristics of the two networks have been described above. For vehicles in the urban area, various real-time services, from safety and emergency services to informational and entertainment applications, are expected to be achieved from the Internet. To guarantee the QoS of vehicular communications, an intelligent network access system (INAS) for service provisioning is necessary, when vehicles are roaming within the coverage of heterogeneous networks. Thus, how to establish the INAS which can provide a seamless ABC vehicular communication based on criteria such as traffic status, network condition and user requirements is the main problem to be studied in this paper. For simplification and without loss of generality, we make following assumptions.

Assumption 1 The geographic area is modeled as a grid-like street layout.

The modeled street pattern consists of a set of M vertical roads intersected with a set of M horizontal roads, as shown in Fig. 4.3. Each line segment of equal length represents a road segment and the region with four road segments around is a street block. The total number of road segments (the road section between any two neighboring intersections) is $G = 2(M - 1)^2$, and each road segment is marked as road segment r ($r = 1, 2, \cdots, G$), from horizontal roads to vertical roads.

Assumption 2 Two types of networks are considered, cellular network and VANET. Cellular network covers the whole region, while VANET covers partially. Additionally, the vehicle-to-vehicle communication is not considered.

Assumption 3 The considered criteria are listed in detail in the following Table 4.1.

For a network, if and only if its received signal strength (RSS) is above a pre-defined threshold, it can be seen as available. Hence, RSS is defined as a triggering factor for the network selection, as well as the change of applications. Additionally, characteristics of networks, bandwidth, delay and cost are defined as influencing

Table 4.1 The considered key criteria

Triggering factors	RSS (availability of the network), application types
Critical factor	Traffic density (terminal mobility)
Influencing factors	Bandwidth, delay, cost

factors. From the aspect of traffic condition, the traffic distribution is seriously uneven. Different traffic density of segments can serious influence on the network selection. Hence, traffic density is defined as the critical factor.

Assumption 4 The price of cellular network is defined as 1 Yuan per MB. For the price of VANET, we use a monthly subscription, that is 10 Yuan per 2 GB for a month.

Since TD-SCDMA standard is proposed by China and used by China Mobile Communication Corp, we take the way China Mobile earn money to price our services. We take a nominal charge for VANET, 10 Yuan per 2 GB for a month.

Assumption 5 In this paper, we just consider the vehicular network selection in a moment.

4.2   Intelligent Network Access System for Real-Time Service Provisioning

In the context of future wireless networks, to achieve a seamless "Always Best Connected" (ABC) vehicular communication, we presents an intelligent network access system (INAS) for vehicular users' real-time services provisioning. The simple block diagram of INAS is shown in Fig. 4.4, which is corresponding to the framework. The network selection mechanism of INAS consists of two components, the network recommendation conducted by the central controller and the automatic network selection conducted by vehicle, respectively. To achieve the goal of network selection based on the aforementioned two components, three phases, *information gathering*, *network selection* and *access execution*, are described in detail.

(1) *Information Gathering*

The *context repository* module is equivalent to the knowledge base which collects the information for network selection from the urban traffic. The data are monitored and updated periodically. *Network repository* holds the network related contexts, such as bandwidth offered by networks, delay for connection establishment and network availability (RSS). *Terminal repository* holds the mobility information of vehicles, like speed, location. *Requirement repository* holds vehicles' preferences which are based on the type of applications and the QoS requirements. *Service repository* stores mainly information like cost of services, services offered.

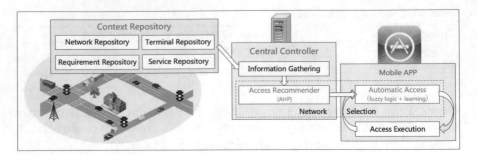

Fig. 4.4 Block diagram of INAS

(2) *Network Selection*

This phase determines whether and how to perform the service provisioning by selecting the most suitable network. Network selection is operated by both central controller and vehicles. Access recommender console is installed in central controller, it provides a guidance of access option in a certain region. The vehicle equipped with an APP acts as automatic access engine, it considers QoS, observes local information, updates local knowledge base, and makes access decision.

(3) *Access Execution*

Result of the network selection phase is executed by the APP in this phase.

Next, we detailed describe the following two algorithms in the aforementioned network selection phase.

- Access recommender console: the network recommendation mechanism.
- Automatic access engine: the automatic selection mechanism.

4.2.1   Tier Model of Vehicular Distribution

As stated in Assumption 1, we model the urban area as a scalable grid with equal length segments and fixed SPs (which can affect the distribution of vehicles). N vehicles are considered in this area with a restricted region called mobility region. As mentioned in former, the distribution of vehicles follows power-law decay from the center of SP to the border of their mobility region. We assume that vehicles' mobility region is partitioned into multiple tiers co-centered at their SPs as shown in Fig. 4.5. Hence, *Tier* model is formed. $Tier(1)$ of the mobility region is collocated with the four road segments around SPs. The adjacent street blocks surrounding $Tier(1)$ form $Tier(2)$, and so on. We denote $Tier(\Lambda)$ as the outermost tier of the mobility region, and $Tier(\beta), \beta \in (1, 2, \cdots, \Lambda)$ to easily expression. For $Tier(\beta)$, it contains $16\beta - 12$ road segments, and vehicle has equal steady-state probability to appear on each road segment. Let π_β denotes the steady-state location probability of

Fig. 4.5 Different tiers centered at a social spot

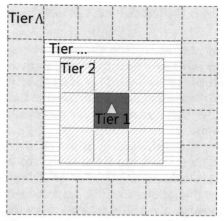

Road segment of Tier 1 ——   Road segment of Tier 2 ······
Road segment of Tier Λ − −   △ Social spot

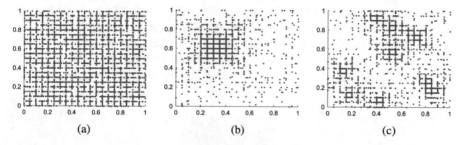

(a)   (b)   (c)

Fig. 4.6 Examples of distribution of vehicles in urban area, in case of N = 2000, M = 21, Λ = 10, and $\gamma = 2$. (**a**) Uniform distribution. (**b**) Distribution with 1 social spot. (**c**) Distribution with 10 social spots

each vehicle on one of segments of $Tier(\beta)$. Due to the exponential fitting curve as analyzed in Sect. 4.1.3, we assume that $\pi_\beta = \beta^{-\gamma}\pi_1$, where γ is the exponent. As the sum of steady-state probability on road segments of the mobility region equals to 1, we have

$$\sum_{\beta=1}^{\Lambda} (16\beta - 12)\pi_\beta = \sum_{\beta=1}^{\Lambda} (16\beta - 12)\beta^{-\gamma}\pi_1 = 1 \qquad (4.1)$$

Thus, the steady-state probability of each road segment is calculated. Figure 4.6 illustrates the distribution of vehicles with different number of SPs. From this figure, we can see that due to the existence of social spots, traffic density is very uneven, and it will serious influence on the network recommendation.

4.2.2 AHP-Based Access Recommender Console

To recommend an "optimum network" to vehicles based on multiple criteria, we introduce AHP developed by Saaty, which is a mean for multi-criteria decision under different conditions. Steps of using AHP are as follows:

Step-I: Modelling the network recommendation problem as a hierarchy which contains the goal, alternatives for reaching the goal, and criteria for evaluating alternatives.

The constructed hierarchy is shown in Fig. 4.7. In the hierarchy, "Optimum Network" is the goal, and it can be realized through selecting an alternative from cellular network and VANET. To reach this goal, traffic density, bandwidth, delay to setup the communication link and cost are defined as criteria. Additionally, RSS and the change of applications act as triggering factors, which will be analyzed in the next section.

Step-II: Establishing priorities among the elements of the hierarchy by making a series of judgments based on pair-wise comparisons of these elements. It is to say that, for each level of the hierarchy, criteria are compared pair-wise according to their influence on a specified criterion in their higher level. The compared results construct a pair-wise comparison matrix in which every element is based on a standardized comparison scale of nine levels as shown in Table 4.2.

Firstly, we construct the pair-wise comparison matrix for the first level in the hierarchy. Let $C = \{C_j | j = 1, 2, \ldots, N_C\}$ be the criteria set of second level.

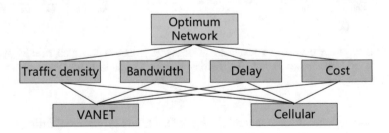

Fig. 4.7 Hierarchy of criteria for "optimum network" selection

Table 4.2 The nine-point scale of pair-wise comparison

Weights	Verbal scale
1	Equal importance of both factors
3	Moderate importance of one factor over another
5	Strong importance of one factor over another
7	Very strong importance of one factor over another
9	Extreme importance of one factor over another
2, 4, 6, 8	Intermediate values between two judgments
Reciprocals	Reflecting dominance of second alternative compared with the first

The result of the pair-wise comparison on the N_C criteria can be summarized in a first level AHP matrix $A = \{a_{ij}\}_{N_C \times N_C}$, in which each element a_{ij} is the quotient weights of criteria, as shown below:

$$A = \begin{bmatrix} a_{11} & a_{12} & \cdots & a_{1N_C} \\ a_{21} & a_{22} & \cdots & a_{2N_C} \\ \vdots & \vdots & \ddots & \vdots \\ a_{N_C1} & \cdots & \cdots & a_{N_CN_C} \end{bmatrix}, a_{ij} = 1 \text{ for } i = j, \text{ and } a_{ij} = 1/a_{ji} \text{ for } a_{ij} \neq 0.$$

(4.2)

The matrix A can be gained based on human knowledge.

Secondly, for each criterion in the second level, N_{net} alternatives of the third level are regarded as criteria. Pair-wise comparison matrices can be constructed according to the influence of third level on the specified criterion C_k, $(k = 1, 2, \ldots, N_C)$ in the second level. Therefore, N_C second level AHP matrices are constructed, defined as $B = \{B_k | k = 1, 2, \ldots, N_C\}$, where $B_k = \{b_{ij}\}_{N_{net} \times N_{net}}$.

The specified characteristics of networks (bandwidth, delay and cost) can be quantified from human knowledge. However, different from network characteristics, traffic density whose influence on networks can not be known, the elements of the second level AHP matrix should be calculated.

Due to the carrier sense with collision avoidance (CSMA/CA) is used as the fundamental access mechanism to the wireless media for the IEEE 802.11p/WAVE, traffic density seriously influence on the probability of vehicles accessing to RSU. An improved protocol, in which the backoff procedure of the IEEE 802.11p is modeled as a p-persistent CSMA/CA proposed in [26]. In the p-persistent CSMA/CA, the backoff interval is based on a geometric distribution with a specific probability of transmission p. Therefore, the probability that a vehicle stays idle when having a busy medium is $1 - p$. For road segment $r \in \{1, 2, \cdots, G\}$, the coverage of an RSU is one-dimensional along road, as shown in Fig. 4.8. The coverage radius of RSU and vehicle are defined as L_R, L_r, respectively. Let the location of RSU be the coordination reference, we define the vehicle whose distance from the reference is d as $vehicle_d$. For $vehicle_d$, whether it has a successful transmission or not is decided by both the channel quality and collisions between two or more vehicles transmitting at the same time. According to [28], we have

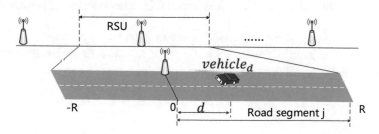

Fig. 4.8 VANET with deployment of RSUs

$$P_{L_R}^{\varsigma}(d) = Q(\frac{10\alpha}{\sigma}\log_{10}\frac{d}{L_R}), \tag{4.3}$$

where $P_{L_R}^{\varsigma}(d)$ is the probability of $vehicle_d$ being directly connected to RSU under channel model ς, which can represent the channel quality of $vehicle_d$; $Q(x)$ is the tail probability of the standard normal distribution, which is the formulation for the log-normal shadowing channel model. α is the path loss exponent, and σ is the variance of a Gaussian random variable. The probability that only $vehicle_d$ has the transmission among all vehicles located at its transmission range is

$$P_d = p(1 - p)^{2L_r\rho_r - 1}, \tag{4.4}$$

where ρ_r is the density of the road r which $vehicle_d$ is located at. Thus, the expectation of the successful transmission probability for vehicles located at road r can be calculated from following equation.

$$E[P_{success}] = \int_0^{L_R} P_{L_R}^{\varsigma}(d) P_d \frac{\pi_r}{L_R} dd, \tag{4.5}$$

where π_r is the probability that vehicles are located at road r. What's more, based on [26], $\frac{1}{p} = \frac{\overline{CW}+1}{2}$, and the CW (contention window size) has been defined in the IEEE 802.11p protocol for different applications.

From the former analysis, the value of density-tolerance for any type of services (*application*) in VANET can be calculated. That is to say, when the density of a road is beyond a threshold $\rho_{application}$ for an application, $E[P_{success}]$ is almost close to zero. For cellular network, it can support traffic density up to traffic jam ρ_{jam} for any type of services. Hence, we can convert the densities of roads to a 9-point scale to form the AHP matrix. The 9-point conversion of density for each network is shown in Eq. (4.6).

$$\rho_{cellular} = 9 - (\rho \times 8)/\rho_{jam}$$
$$\rho_{VANET} = 9 - (\rho \times 8)/\rho_{application} \tag{4.6}$$

Step-III: The pair-wise comparison matrix should satisfy transitive preference and strength relations, it is necessary to check its consistency. When the comparison matrix passes the consistency test, it is acceptable, otherwise its judgments should be revised.

The consistency test includes two important calculation processes for Consistency Index (*C.I.*) and Consistency Ratio (*C.R.*), whose formulas are shown in following equation,

$$C.I. = (\lambda_{\max} - n)/(n - 1); \quad C.R. = C.I./R.I. \tag{4.7}$$

where n represents the size of the pair-wise comparison matrix, and λ_{max} denotes the maximum Eigen value of the matrix; $R.I.$ (Random Index) is a random consistency index. If $C.R. \leq 0.1$, the inconsistency is acceptable, otherwise the judgments should be revised.

Step-IV: Ranking the priority vector of each level's alternatives.

If the constructed matrix has passed the consistency test, the priority vector (V) can be calculated, which represents the priority of the compared factors in the same level for deciding the ranking of importance weights on a criterion in the higher level. V can be gain from normalizing the eigenvector corresponding to the maximum eigenvalues of the comparison matrix.

$$AV = \lambda_{max}V, \ V = [V_1, V_2, \ldots, V_n]' \tag{4.8}$$

Thus, priority vectors of A and B_k can be got, they are $V_A = [V_{A_1}, V_{A_2}, \ldots, V_{A_{N_C}}]'$ and $V_{B_k} = [V_{B_{k1}}, \ldots, V_{B_{kN_{net}}}]'$.

Step-V: Synthesize these priority vectors to construct an overall priority vector, and the final priorities of the alternatives for the goal can be yield.

The final priorities of the alternatives (V_i) are determined by multiplying the priority vector of the criteria found from first level AHP matrix (V_A) by the priorities found from each second level AHP matrix (V_{B_k}) for the goal.

$$V_i = \sum_{k=1}^{N_c} V_{Bki} V_{Ak}, \ i = 1, \ldots, N_{net}. \tag{4.9}$$

The final priorities of alternative networks for the "Optimum Network" can be got through the above five steps. Traffic density is the critical factor and reflects the feature of vehicles' mobility. It is evident from Eq. (4.6) that if traffic density is more than $\rho_{application}$, VANET is theoretically considered as not available and network priority will depend only on the influence factors. However, the unavailability does not mean all vehicles can not access to VANET, it just means vehicles can not access to VANET with a bigger probability. So, it is necessary to design vehicle's automatic access mechanism under controller's recommendation.

4.2.3 Learning-Based Automatic Access Engine

The engine of *Mobile APP* operators in an automatic process shown in Fig. 4.9.

The QoS requirements of various applications are registered with local observation of speed, the sensed RSS of networks, and the pushed message of access recommender. The access option can be decided by intelligent analysis methods with combinations of the registered information, the current achieved QoS and the knowledge in the past time. It is noted that the knowledge base is defined as

$Q\langle speed, RSSc, RSSv, application, option, QoS \mid recommendation\rangle,$

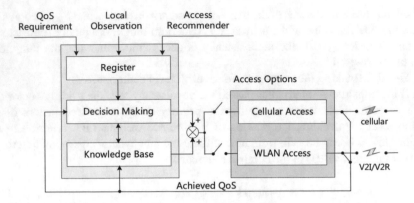

Fig. 4.9 Automatic decision making process

which is updated by the new achieved QoS with a proper frequency. Thus, the trustworthiness of access recommender can be adapted according to local observation and achieved QoS. The adapted process is realized through machine learning. After the training of history information, a mapping relationship (Γ) on *speed, RSSc, RSSv, application, recommendation, option* and *QoS* can be achieved.

$$QoS = \Gamma \ (speed, RSSc, RSSv, application, recommendation, option).$$

Then, vehicle can make access decision through the mapping relationship when a new network selection is required.

4.2.3.1 Utility Function

QoS refers to the satisfaction that a service provides to a user. Multiple attributes, $\langle data\ rate, delay, cost \rangle$, make contributions to vehicles' QoS in our problem. For *data rate*, the higher preference relation is in favor of the higher value. Conversely, *delay* and *cost* prefer lower value for a better QoS. Given an attribute, its utility can be represented based on utility function. A lot of studies have analysed the utility functions of different attributes [15, 27]. We use sigmoidal, linear and logarithmic functions to evaluate the utility of the considered attributes as shown in the following functions.

$$u_1(x_1) = (x_1/x_m)^{a_1}/(1 + (x_1/x_{1m})^{a_1}) \tag{4.10}$$

$$u_2(x_2) = 1 - a_2 x_2 \tag{4.11}$$

$$u_3(x_3) = 1 - \ln(1 + a_3 x_3)/\ln(1 + x_3) \tag{4.12}$$

where $u_i(x_i)$, ($i = 1, 2, 3$) represents the utility function of *data rate, delay, cost*, respectively. a_i and x_m are parameters depending on different functions.

Due to multiple attributes are considered to calculate QoS, the utilities of all attributes should be combined as a total utility. A valid form of total utility should satisfy the following four requirements [27]:

$$\begin{cases} \frac{\partial U(\mathbf{x})}{\partial u_i} \geq 0 \\ \text{sign}(\frac{\partial U(\mathbf{x})}{\partial x_i}) = \text{sign}(u_i'(x_i)) \\ \lim_{u_i \to 0} U = 0, \forall i = 1, \cdots, N_A \\ \lim_{u_1, \cdots, u_{N_A} \to 1} U = 1 \end{cases} \tag{4.13}$$

where $\mathbf{x}$ is the attributes vector and N_A is the number of attributes. The first requirement means that the aggregate utility should increase for increasing elementary utility. The second requirement means that the monotonicity of aggregate utility function and elementary utility functions should be consistent. The third condition ensures that elementary functions are independent. If the utility of one attribute is zero, it must result in the zero value of the aggregate utility. The forth condition reflects the fact that if all elementary utilities are perfectly satisfied, the aggregate utility should be too.

Given an attribute vector $\mathbf{x}$ and the associated preference vector $\boldsymbol{\omega}$, the suitable aggregate multi-attributes utility function is formulated as:

$$U(\mathbf{x}) = \prod_{i=1}^{N_A} [u_i(x_i)]^{\omega_i} \tag{4.14}$$

where ω_i is the preference weight of attribute i ($\sum_{i=1}^{N_A} \omega_i = 1$). $u_i(x_i)$ is the elementary utility function of attribute i.

Additionally, the *RSS* can directly impact on the *data rate*. It has been described in [28], the vehicle and RSU can establish a direct connection if the RSS at the vehicle is greater than or equal to a certain threshold, and the directly connection probability can decide the data rate vehicle gained.

Thus, different utilities for different vehicles with different local information, preferences and services can be achieved, so as the QoS of vehicles.

4.2.3.2 Function Approximation

In order to achieve the mapping relationship (Γ) and a proper trustworthiness of access recommender through machine learning, we use fuzzy network to train the historical information which can decrease the computational complexity and ensure the training accuracy.

At first, we use fuzzy rules to represent the relation between the achieved $\langle QoS \rangle$ and its influence factors $\langle speed, RSSc, RSSv, application, recommendation, option \rangle$

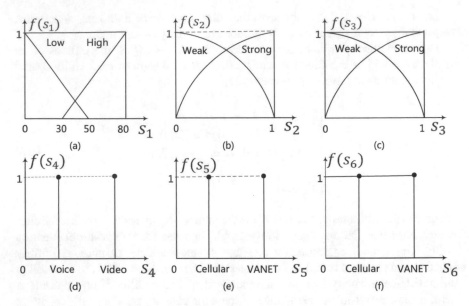

Fig. 4.10 The membership functions

which can be abbreviated as $\langle s_1, \ldots, s_6 \rangle$. In fuzzy theory, a rulebase is the function Γ that maps the input vector into a scalar output.

To have a general description about our problem, we assume that there are n input variables, $\langle s_1, \ldots, s_n \rangle$ which are premise variables. The achieved QoS is defined as the output variable. Each of premise variables consists of two fuzzy sets. In specific, s_1 is divided into *Low* and *High*, both s_2 and s_3 are divided into *Weak* and *Strong*, we give s_4 two specific definitions, which is divided into *Voice* and *Video*, both s_5 and *VANET* are divided into *Cellular* and *VANET*. The membership function $f(s_i)$ of each variable is shown in Fig. 4.10. s_1 is in the range of $0 \sim 80$ km/h, its membership function is represented by linear. We transform s_2 and s_3 into the connection probability which are in range of $0 \sim 1$, their membership function are represented by normal. s_4, s_5, and s_6 are solid, we use points to represent them. Hence, we get the following 64 fuzzy rules:

- Rule 1: If s_1 is *Low*, s_2 is *Weak*, s_3 is *Weak*, s_4 is *Voice*, s_5 is *Cellular*, s_6 is *Cellular*, then $QoS = o'_1$;
- Rule 2: If s_1 is *Low*, s_2 is *Weak*, s_3 is *Weak*, s_4 is *Voice*, s_5 is *Cellular*, s_6 is *VANET*, then $QoS = o'_2$;
- $\ldots$;
- Rule 64: If s_1 is *High*, s_2 is *Strong*, s_3 is *Strong*, s_4 is *Video*, s_5 is *VANET*, s_6 is *VANET*, then $QoS = o'_{64}$.

These fuzzy rules establish the mapping Γ from fuzzy logic conditions to output QoS. For example, the second rule is interpreted as *"If a vehicle's speed is low, both of its RSS from cellular network and VANET are weak, its application is voice and*

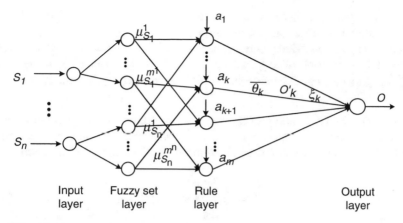

Fig. 4.11 Multi-layer feed forward fuzzy network

the access recommender is cellular but its access option is VANET, then the QoS it can get is o'_2".

4.2.3.3 Network Selection Based on Learning

In order to train the mapping function Γ between $\langle s_1, \ldots, s_n \rangle$ and $\langle QoS \rangle$, fuzzy network trained with a back-propagation algorithm is used in our work. The network consists of four layers which are input layer, fuzzy set layer, rule layer and output layer as shown in Fig. 4.11.

1. Input layer. Each node in this layer represents one premise variable, the total number of nodes is recorded as n. The inputs of the first layer are $s_1, \ldots, s_n$, the outputs are the same as the inputs.
2. Fuzzy set layer. Each node in this layer represents a fuzzy set, each premise variable s_i is divided into m^i sets. The total number of nodes is $\sum_{i=1}^{n} m^i$. Then, the membership grade of each premise variable $\mu_{s_i}^j, j = 1, \ldots, m^i$ belongs to each set can be calculated through membership functions $f(s_i)$. The inputs of the second layer are $s_1, \ldots, s_n$, the outputs are $\mu_{s_i}^j, j = 1, \ldots, m^i$.
3. Rule layer. Each node of this layer represents a rule, the total number of nodes is defined as m, $m = \prod_{i=1}^{n} m^i$. The fitness of each rule should be calculated with the membership function of each variable. For the kth rule, its fitness is $\theta_k = \prod_{i=1}^{n} \mu_{s_i}^j, j \in 1, \ldots, m^i (k = 1, \ldots, m)$. the mean value of the fitness is $\overline{\theta_k} = \frac{\theta_k}{\sum_{k=1}^{m} \theta_k}$. Each rule's output is o'_k. Here, we add an output threshold coefficient a_k of each corresponding rule in order to train the network more precisely. The inputs of the third layer are $\mu_{s_i}^j$ and a_k, the outputs are $\overline{\theta_k}$ and o'_k.

4. Output layer. The node of this layer represents the output. To realize the defuzzification of the output, we need to calculate the precise output o when the inputs of the whole network are $s_1, \ldots, s_n$, combined with the weight coefficients of each rule, ξ_k. Then, the precise output can be calculated, $o = \sum_{k=1}^{m} \xi_k \overline{\theta_k} o'_k + a_k$. The inputs of the forth layer are $a_k, \overline{\theta_k}, o'_k$ and ξ_k, the output is o.

The supervised adoption process varies the threshold and weight coefficients to minimize the sum of the squared differences between the computed and the required output values. This is accomplished by minimizing the error criterion function E.

$$E = \frac{1}{2} \sum_{i=1}^{N_T} (o_i - y_i)^2 \tag{4.15}$$

where N_T is the number of training samples, y_i is the required output, o_i is the computed output.

In back-propagation algorithm, weight and threshold coefficients are adjusted by the following equations:

$$\Delta \xi_k = -\eta \frac{\partial E}{\partial \xi_k}, \Delta a_k = -\eta \frac{\partial E}{\partial a_k} \tag{4.16}$$

where η is the rate of learning ($\eta > 0$).

Thus, vehicles can get themselves rulebase's functions Γ, which map their input parameters, $\langle s_1, \ldots, s_6 \rangle$, into the output QoS through training network with the former learning method. For a vehicle with local information, $\langle speed, RSSc, RSSv, application \rangle$, it can predict the $\langle QoS \rangle$ gained from networks, $\langle option \rangle$, through its well trained rulebase's function under the access recommender information $\langle recommendation \rangle$. When it has the knowledge of different types of QoS, it can choose whether or not to adopt the recommendation of access recommender and select a more proper network.

4.3   Performance Evaluation

4.3.1   Simulation Methodology

In this section, we evaluate the performance of INAS using simulations over MATLAB. We simulate a grid-like city area of 10×10 km which are composed by $M = 20$ vertical and horizontal streets intersected with each other, the length of each road segment is 500 m. 5 social spots are considered, and 20,000 vehicles with the mobility following social-based model described in Sect. 4.1.3 are moving in the simulation. The outmost tier mobility region of vehicles is set to be 10, and the exponent $\gamma = 2$. Then, the snapshot of the vehicle distribution is shown in Fig. 4.12.

Fig. 4.12 Distribution of vehicles

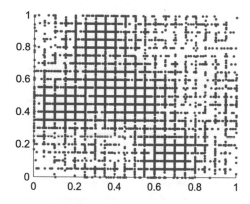

Table 4.3 1st level AHP comparison matrix

(a) Voice service

Optimum network	Traffic density	Bandwidth	Delay	Cost	Priority vector
Traffic density	1	5	3	7	0.5558
Bandwidth	1/5	1	1/3	5	0.1364
Delay	1/3	3	1	5	0.2589
Cost	1/7	1/5	1/5	1	0.0489
	$C.I. = 0.0801, C.R. = 0.0890$				

(b) Video service

Optimum network	Traffic density	Bandwidth	Delay	Cost	Priority vector
Traffic density	1	1/7	1/5	1/3	0.0553
Bandwidth	7	1	3	5	0.5650
Delay	5	1/3	1	3	0.2622
Cost	3	1/5	1/3	1	0.1175
	$C.I. = 0.0801, C.R. = 0.0890$				

As defined in Assumption 5, two different categories of the applications, voice and video communications, are simulated. In general, a vehicle on average spending 3 min on voice service and 5 min on video service. The data flow rate is 0.6 kb/s and 5 Mb/s for voice and video service, respectively. Due to the method of AHP is used to solve the network recommendation problem, the pair-wise comparison matrices should be constructed. Table 4.3 shows the first level AHP comparison matrix, which is used for deciding priority among the following criteria to advise vehicle with a network to access (cellular or VANET): traffic density, bandwidth, delay, and cost. In Table 4.3, (a) and (b) present the matrices for voice and video applications, respectively. For example, vehicles have a more frequent requirement for voice communication. We define traffic density as the most importance factor among the four criteria. The delay for connection establishment should be short for voice communication, the bandwidth and cost requirements are not high. For

Table 4.4 2nd level AHP comparison matrices

(a)

Bandwidth	Cellular	VANET
Cellular	1	1/7
VANET	7	1
	$C.I. = 0$	

(b)

Delay	Cellular	VANET
Cellular	1	1/5
VANET	5	1
	$C.I. = 0$	

(c)

Cost	Cellular	VANET
Cellular	1	1/7
VANET	7	1
	$C.I. = 0$	

(d)

Density	Cellular	VANET
Cellular	1	$\frac{\rho_{cellular}}{\rho_{VANET}}$
VANET	$\frac{\rho_{cellular}}{\rho_{VANET}}$	1
	$C.I. = 0$	

video communication, it pays more attention to bandwidth and delay than cost. Additionally, we can observe that the value of $C.R.$ is less than 0.1 from both of the former two tables, which means the constructed pair-wise comparison matrices have passed the consistency test. Thus, the first level of priority vectors can be got.

The second level AHP comparison matrices are introduced to decide priority of each network, which are shown in Table 4.4. The bandwidth and cost feature of each network have been discussed before, we construct corresponding comparison matrices as shown in Table 4.4a, c. According to [29], we define the delay of cellular and VANET as 270 ms and 30 ms, respectively. The comparison matrix is shown in Table 4.4b.

For the traffic density, its second level pair-wise comparison matrix is formed as Table 4.4d. The simulation result about the values of density-tolerance for the two applications, $\rho_{application}$, shows that the successful transmission probability is nearly zero when traffic density is 0.04 for voice service, and 0.06 for video service.

4.3.2   Access Recommender Console Results

In this subsection, we simulate results of access recommender console based on the constructed AHP matrices and the distribution of vehicles as shown in Fig. 4.12. For voice and video services, the network recommendation index are shown in Figs. 4.13 and 4.14, where (a) and (b) represent the recommendation index of cellular network and VANET, respectively. Due to the recommender console takes road as unit, we integrate the four road segments' recommendation indexes around a street block into one to intuitively describe the index. Hence, the index just represents the trend of network recommended, so for a recommendation index in the figure, the sum of cellular network and VANET is not equal to one. However, we can achieve that the recommendation index of cellular network is higher than that of VANET when the traffic density is higher as shown in Figs. 4.13 and 4.14. It indicates that traffic density has a larger influence on the network recommendation, and it is very crucial

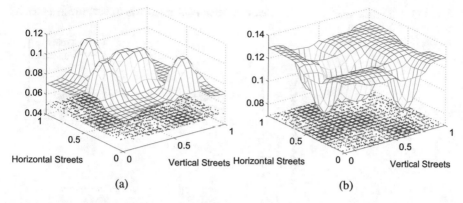

Fig. 4.13 Recommendation index of cellular network and VANET for voice service. (**a**) Cellular network. (**b**) VANET

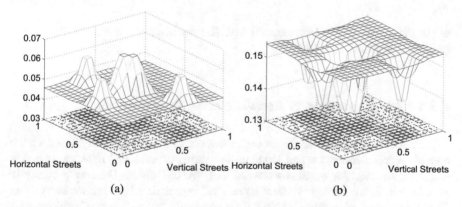

Fig. 4.14 Recommendation index of cellular network and VANET for video service. (**a**) Cellular network. (**b**) VANET

in network selection. This mainly results from the different priorities of the four factors for different services.

4.3.3 Fuzzy and Learning Effects

The access recommender console is designed based on AHP, and it can recommend vehicles which network to access to achieve better QoS. However, for a vehicle, the network access decision should be made by itself. In order to achieve the automatic access, the knowledge base is trained. In the simulation, 50, 50 and 50 independent samples have been used for training, validation and testing to check the capability

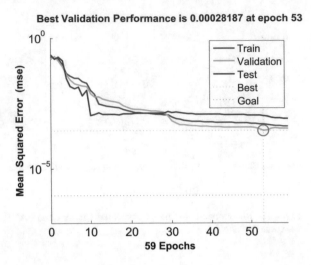

Fig. 4.15 Performance of training, validation and testing of training network

of the trained fuzzy network, respectively. The performance of the network during training is shown in Fig. 4.15.

4.3.4   Automatic Access Results Under Recommender

After vehicles can make decisions on network access autonomously based on the console recommender, we can simulate the status of vehicles' network selection in the urban area. As we have assumed that VANET covers the area partially in Assumption 2, we randomly select some road segments which are far away from those social spots as without VANET coverage roads. We set a part of vehicles with voice requirements and others with video requirements. Figures 4.16 and 4.17 show the status of vehicles' network selection based on vehicular random behaviors and the application of INAS in a moment. For each figure, (a) and (b) represent the status of network selection for voice service and video service, where red circles mean vehicles access to cellular network, green points mean vehicles access to VANET and blue circles mean vehicles without network to access. Comparing Fig. 4.17 with Fig. 4.16, vehicles are in favor of selecting cellular network with higher traffic density, and the trend is more obvious for voice service than video service with the application of INAS. Two main reasons may account for this phenomenon. On one hand, it is because that the size of video service is much more than that of voice service, so the expense of video service is higher. On the other hand, more requirements for voice service results in lower endurance capacity in VANET than video service. These two reasons result in the existence of those blue circles. According to statistics, for vehicles with voice and video requirement, there are about 18% and 10% among them can not access to network under random selection scheme, respectively. However, there are correspondingly about 2% and 5% vehicles

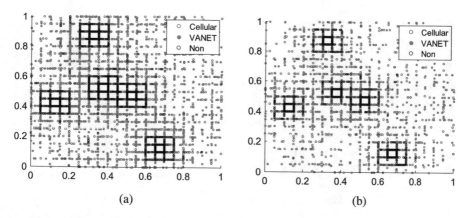

Fig. 4.16 Status of vehicles' network selection with random. (**a**) Voice service. (**b**) Video service

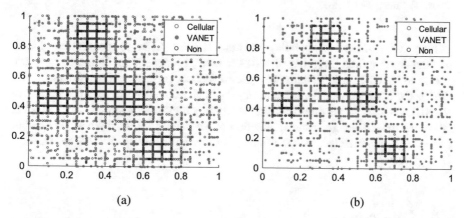

Fig. 4.17 Status of vehicles' network selection with INAS. (**a**) Voice service. (**b**) Video service

can not access to network under the INAS scheme, respectively. It demonstrates that INAS does better network selection compared with the random behavior.

After vehicles access to corresponding network, the total QoS of them can be calculated from the aggregate utility function. Figure 4.18 shows the different proportion of each type of scheme in the sum of the three types of network selection schemes. "All cellular" means the situation without VANET, all vehicles assess to cellular network to gain services. "Random" is the scheme that vehicles select a network to access with a random behavior as analysed earlier. "INAS" is the scheme we proposed. As VANET uses a monthly subscription and the monthly rent is deducted at the first time of use of VANET, the advantage of VANET is more obvious with multiple use. Figure 4.18a, b show the proportion of each type of scheme when vehicles have one time service requirement and 30 times service requirement, respectively. The weights of attributes in the process of calculating the total utility function is gained from the first level of AHP matrix, the normalization

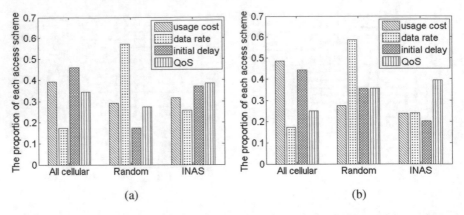

Fig. 4.18 Performance of QoS with different network selection scheme and different service times. (**a**) 1 time. (**b**) 30 times

of the priority of bandwidth, delay and cost. For Fig. 4.18a, the proportion of QoS of "Random" is less than that of "All cellular", because many vehicles have not access to VANET. Additionally, In spite of the largest data rate of "Random", its QoS is the least because there are too many vehicles with voice service connect to VANET. As a result, the large data rate can barely improve the QoS of voice service. However, the advantage of the application of VANET is reflected when the number of service requirements is larger as shown in Fig. 4.18b, and also INAS has the best QoS performance.

4.4   Conclusion

We argue that with the foreseeable surge deployments of vehicular bourn systems (iOS and Android in cars), it can be envisioned that traditional cellular networks will soon encounter an explosive growth of vehicular traffic, which makes an efficient offloading scheme necessary. To address this issue, we have presented a vehicular networking system which enables efficient offloading for real-time traffic to cellular networks. Our proposal is composed of two parts. From the network perspective, we have developed an access recommender console deployed in the central controller; the console provides efficient guidance to vehicles for network accessing. From the perspective of individual vehicles, an automatic access engine is developed to install on distributed vehicles, which guide vehicles to make decisions on network selection based on combinations of the access recommender, the local observation and the knowledge accumulated in the past. We have conducted extensive simulations, and shown that the proposed network selection scheme provides high network resource utilization and guaranteed QoS to vehicles.

References

1. S. Greengard, *Automotive systems get smarter*, Commun. Acm, vol. 58, no. 10, pp. 18–20, 2015.
2. M. G. Demissie, G. Correia and C. Bento, *Exploring Cellular Network Handover Information for Urban Mobility Analysis*, IEEE J. Transport Geography, vol. 31, pp. 164–170, 2013.
3. H. Dong, X. Ding, Y. Shi, L. Jia, Y. Qin, and L. Chu, *Urban traffic commuting analysis based on mobile phone data*, IEEE 17th International Conference on Intelligent Transportation Systems, pp. 611–616, 2014.
4. *Wireless LAN Medium Access Control (MAC) and Physical Layer (PHY) Specifications Amendent: Wireless Access in Vehicular Environment*, IEEE Std, P802.11p, Jul, 15, 2010.
5. S. Al-Sultan, M. M. Al-Doori, A. H. Al-Bayatti, and H. Zedan, *A comprehensive survey on vehicular ad hoc network*, J. Netw. Comput. Appl., vol. 37, pp. 380–392, 2014.
6. E. C. Eze, S. Zhang, and E. Liu, *Vehicular ad hoc networks (vanets): Current state, challenges, potentials and way forward*, International Conference on Automation and Computing (ICAC), pp. 176–181, 2014.
7. D. Tian, J. Zhou, Y. Wang, Y. Lu, H. Xia, and Z. Yi, *A dynamic and self-adaptive network selection method for multimode communications in heterogeneous vehicular telematics*, IEEE Trans. Intell. Transp. Syst., vol. 16, no. 6, pp. 3033–3049, 2015.
8. O. Kaiwartya, A. H. Abdullah, Y. Cao, A. Altameem, M. Prasad, C.-T. Lin, and X. Liu, *Internet of vehicles: Motivation, layered architecture, network model, challenges, and future aspects*, IEEE Access, vol. 4, pp. 5356–5373, 2016.
9. J. Marquez-Barja, C. T. Catafate, J. C. Cano and P. Manzoni, *An Overview of Vertical Handover Techniques: Algorithms, Protocols and Tools*, Computer Communications, vol. 34, no. 8, pp. 985–997, 2011.
10. M. Kassar, B. Kervella and G. Pujolle, *An Overview of Vertical Handover Decision Strategies in Heterogeneous Wireless Networks*, Computer Communications, vol. 31, no. 10, pp. 2607–2620, 2008.
11. R. K. Goyal, S. Kaushal, and S. Vaidyanathan, *Fuzzy ahp for control of data transmission by network selection in heterogeneous wireless networks*,International Journal of Control Theory and Applications, vol. 9, no. 1, pp. 133–140, 2016.
12. K. Xu, K. Wang, R. Amin, J. Martin, and R. Izard, "A fast cloud-based network selection scheme using coalition formation games in vehicular networks," *IEEE Trans. Veh. Technol.*, vol. 64, no. 11, pp. 5327–5339, 2015.
13. E. Stevens-Navarro, Y. Lin, and V. W. S. Wong, *An mdp-based vertical handoff decision algorithm for heterogeneous wireless networks*, IEEE Trans. Veh. Technol., vol. 57, no. 2, pp. 1243–1254, 2008.
14. Q. Wu, Z. Du, P. Yang, Y. Yao, and J. Wang, *Traffic-aware online network selection in heterogeneous wireless networks*, IEEE Trans. Veh. Technol., vol. 65, no. 1, pp. 381–397, 2016.
15. L. Wang and G. S. Kuo, *Mathematical Modeling for Network Selection in Heterogeneous Wireless Networks-A Tutorial*, IEEE Communications Surveys & Tutorials, vol. 15, no. 1, pp. 271–292, April. 2013.
16. A. Calvagna and G. Di Modica, *A user-centric analysis of vertical handovers*, Proceedings of the 2nd ACM international workshop on Wireless mobile applications and services on WLAN hotspots. ACM, 2004: 137–146.
17. F. Bari and V. C. Leung, *Automated Network Selection in a Heterogeneous Wireless Network Environment*, IEEE Network, vol. 21, no. 1, pp. 34–40, 2007.
18. J. Hou and D. O'brien, *Vertical handover-decision-making algorithm using fuzzy logic for the integrated Radio-and-OW system*, IEEE Transactions on Wireless Communications, vol. 5, no. 1, pp. 176–185, 2006.
19. T. Ahmed, K. Kyamakya and M. Ludwig, *A Context-Aware Vertical Handover Decision Algorithm for Multimode Mobile Terminals and its Performance*, Proceedings of the IEEE/ACM Euro American Conference on Telematics and Information Systems, pp. 19–28, 2006.

20. L. S. Wang and G. S. Kuo, *Dynamics of Network Selection in Heterogeneous Wireless Networks: An Evolutionary Game Approach*, IEEE trans. Vehicular Technology, vol. 58, no. 4, pp. 2008–2017, 2009.
21. A. Hasib and A. O. Fapojuwo, *Cross-layer Radio Resource Management in Integrated WWAN and WLAN Networks*, Computer Networks, vol. 54, no. 3, pp. 341–356, 2010.
22. G. Kousalya, P. Narayanasamy, J. H. Park, et al, *Predictive handoff mechanism with real-time mobility tracking in a campus wide wireless network considering ITS*, Computer Communications, vol. 31, no. 12, pp. 2781–2789, 2008.
23. S. Lee, K. Sriram, K. Kim, Y. H. Kim and N. Golmie, *Vertical Handoff Decision Algorithms for Providing Optimized Performance in Heterogeneous Wireless Networks*, IEEE trans. Vehicular Technology, vol. 58, no. 2, pp. 865–881, 2009.
24. S. Dhar, A. Ray and R. Bera, *Cognitive Vertical Handover Engine for Vehicular Communication*, Peer-to-Peer Networking and Applications, vol. 6, no. 3, pp. 305–324, 2013.
25. N. Lu and X. Shen, *Capacity Analysis of Vehicular Communication Networks*, Springer, 2013.
26. W. Alasmary and W. Zhuang, *Mobility Impact in IEEE 802.11p Infrastructureless Vehicular Networks*, Ad Hoc Networks, vol. 10, no. 2, pp. 222–230, 2012.
27. Q. T. Nguyen-Vuong, Y. Ghamri-Doudane and N. Agoulmine, *On Utility Models for Access Network Selection in Wireless Heterogeneous Networks*, IEEE NOMS, pp. 144–151, 2008.
28. S. C. Ng, W. Zhang, Y. Zhang, Y. Yang and G. Mao, *Analysis of Access and Connectivity Probabilities in Vehicular Relay Networks*, IEEE J. Sel. Areas Commun., vol. 29, no. 1, pp. 140–150, 2011.
29. J. N. Cao and C. S. Zhang, *Seamless and Secure Communications over Heterogeneous Wireless Networks*, Springer, 2014.

Chapter 5
UAV Relay in VANETs Against Smart Jamming with Reinforcement Learning

Frequency hopping-based anti-jamming techniques are not always applicable in VANETs due to the high mobility of OBUs and the large-scale network topology. In this chapter, we use unmanned aerial vehicles (UAVs) to relay the message of an OBU and improve the communication performance of VANETs against smart jammers that observe the ongoing OBU and UAV communication status and even induce the UAV to use a specific relay strategy and then attack it accordingly. More specifically, the UAV relays the OBU message to another RSU with a better radio transmission condition if the serving RSU is heavily jammed or interfered.

Game theory has been used to study the anti-jamming power control in wireless networks. However, to our best knowledge, the jamming resistance in the UAV-aided VANETs is still an open problem. In this chapter, the interactions between a UAV and a smart jammer are formulated as an anti-jamming UAV relay game, in which the UAV decides whether or not to relay the OBU message to another RSU, and the jammer observes the UAV and the VANET strategy and chooses the jamming power accordingly. The Nash equilibria of the game are derived and the conditions that each NE exists are provided to disclose how the VANET transmission, the jamming model and the UAV channel model impact the BER of the OBU message in the UAV-aided VANET against jamming. A hotbooting policy hill climbing (PHC)-based UAV relay strategy is proposed to help the VANET resist jamming in the dynamic game without being aware of the VANET model and the jamming model. Simulation results show that the proposed relay strategy can efficiently reduce the bit error rate of the OBU message and thus increase the utility of the VANET compared with a Q-learning based scheme.

5.1 UAV-Aided VANET Communication

Vehicular ad-hoc networks (VANETs) support vehicle-to-vehicle communications and vehicle-to-infrastructure communications to improve the transmission security,

© Springer Nature Switzerland AG 2019
L. Xiao et al., *Learning-based VANET Communication and Security Techniques*, Wireless Networks, https://doi.org/10.1007/978-3-030-01731-6_5

help build unmanned-driving, and support booming applications of onboard units (OBUs) [1]. The high mobility of OBUs and the large-scale dynamic network with fixed roadside units (RSUs) make the VANET vulnerable to jamming [2]. A jammer sends faked or replayed signals and aims to block the ongoing transmissions between OBUs and the serving RSUs. By applying smart radio devices to observe the ongoing VANET communication and evaluate the underlying policy, a smart jammer not only has flexible control over the jamming frequencies and signal strengths but also induces the VANET to use a specific communication strategy and then attacks it accordingly.

The anti-jamming communication of VANETs can be significantly improved by using unmanned aerial vehicles (UAVs) to relay the OBU message. Being faster to deploy, UAVs generally have better channel states due to the line-of-sight (LOS) links and smaller path-loss exponents [3, 4] when they communicate with OBUs and RSUs, compared with the serving RSUs at a fixed location on the ground that might be severely blocked by a smart jammer. Therefore, UAVs help relay the OBU message to improve the signal-to-interference-plus-noise-ratio (SINR) of the OBU signals, and thus reduce the bit-error-rate (BER) of the OBU message, especially if the serving RSUs are blocked by jammers and/or interference.

UAVs have been used to relay mobile messages to improve communication performance in wireless networks [3, 5–10]. The optimization of multi-antenna UAVs and mobile ground terminals in [5] improves the uplink sum rate in a wireless relay network. The relay scheme as proposed in [6] uses UAVs to maximize the capacity of a wireless relay network. The UAV-aided intrusion detection scheme presented in [3] uses UAVs to relay the alarm messages regarding lethal attacks of vehicles to improve the detection accuracy and reduce the energy consumption in vehicular networks. The iterative UAV relay algorithm proposed in [7] optimizes the power control and relay trajectory to improve the throughput of mobile networks. The UAV-enabled mobile relaying system as investigated in [8] uses the difference-of-concave program to optimize the transmit power of the mobile device and the relay, and maximize the secrecy data rate.

The UAV placement strategy as developed in [9] enhances the coverage of public safety communications. The UAV-aided sensor deployment in [10] improves the localization and navigation to monitor post-disaster areas. The field tests in [11] show the impact of the UAV altitude on the communication quality in autonomous vehicles. The UAV-aided wireless sensor network as investigated in [12] can reduce the packet loss and power consumption of the network against node failures. The UAV-assisted data gathering system as developed in [13] reduces the required execution time and the energy consumption in wireless sensor networks. The tradeoff between the coverage and the time required to cover the entire target area of the UAV is studied in [14] to determine the number of stop points of the UAV in an underlaid device-to-device communication network.

5.2 UAV Relay Anti-jamming Transmission Game

In this section, we formulate an anti-jamming UAV relay stochastic game, in which the UAV decides whether or not to relay an OBU message to another RSU, and the smart jammer chooses its jamming power under the time-variant random channel power gains, which can be modeled as a Markov chain [15, 16]. The Nash equilibria (NEs) of the game are derived and the conditions that each NE exists are provided to disclose how the VANET transmission, the jamming model and the UAV channel model impact the BER of the OBU message in the UAV-aided VANET against jamming.

5.2.1 Network Model

We consider an OBU that moves along the road at a speed denoted by $v^{(k)} \in [0, V]$ at time slot k, where V is the maximum speed and time is partitioned into slots of a constant duration. The OBU aims to send a message to a server via several RSUs and a UAV in a time slot, as illustrated in Fig. 5.1. The RSUs at fixed locations are connected via fibers with each other and the server. Equipped with sensors such as cameras and a global positioning system receiver, the OBU gathers the sensing information and sends a message to the server via the serving RSU denoted by RSU_1. We assume that both the UAV and RSU_1 receive the message from the OBU and then the UAV decides whether to connect to the server via RSU_2 afterwards in the time slot. For simplicity, the constant channel power gains are assumed to be constant in each time slot.

Let $\mathbf{d}^{(k)} = [d_{OR}^{(k)}, d_{OU}^{(k)}, d_{JR}^{(k)}, d_{JU}^{(k)}, d_{UR}^{(k)}]$ denote the topology vector of the network at time slot k, where $d_{OR}^{(k)}$ denotes the distance between the OBU and RSU_1, $d_{OU}^{(k)}$ is the distance between the OBU and the UAV, $d_{JR}^{(k)}$ corresponds to the distance of the jammer-RSU_1 link, $d_{JU}^{(k)}$ is the distance between the jammer and the UAV,

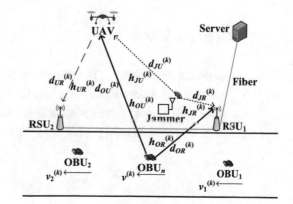

Fig. 5.1 Illustration of a UAV-aided VANET, in which the OBU moving with speed $v^{(k)}$ sends a message at time slot k to a server via the serving RSU (RSU_1) and the UAV that is connected with RSU_2 and is less affected by the jammer

and $d_{UR}^{(k)}$ is the distance between the UAV and RSU$_2$. The distances $d_{OU}^{(k)}$ and $d_{OR}^{(k)}$ depend on the speed of the OBU at time slot k. The OBU sends a message to a server via RSU$_1$ and the UAV that is connected with RSU$_2$ at time slot k with a fixed transmit power $P^{(k)}$. The SINR of the signals received by RSU$_1$ (UAV) sent from the OBU is denoted by $\rho_{OR}^{(k)}$ ($\rho_{OU}^{(k)}$), and the SINR of the signals received by RSU$_2$ sent from the UAV at time slot k is denoted by $\rho_{UR}^{(k)}$. The BER of a signal denoted by $p_e(\rho)$ depends on the SINR per bit, ρ, of the received signal for a given modulation mode. The BER of the OBU message at time slot k denoted by $P_e^{(k)}$ depends on the OBU-RSU$_1$ link, the OBU-UAV link and the UAV-RSU$_2$ link, and is given by

$$P_e^{(k)} = \min\left(p_e\left(\rho_{OR}^{(k)}\right), p_e\left(\min\left(\rho_{OU}^{(k)}, \rho_{UR}^{(k)}\right)\right)\right). \tag{5.1}$$

The BER of the message depends on the minimum of the BER of the OBU-RSU$_1$ signal as shown in the first term, and the BER of the weaker signal of the OBU-UAV link and the UAV-RSU$_2$ link at that time as represented in the second term. According to the channel quality and the BER of the OBU message, the UAV decides whether or not to relay the OBU message to RSU$_2$, which is denoted by $x \in \mathbf{A} = \{0, 1\}$, where $\mathbf{A}$ is the feasible action set of the UAV. The UAV relays the OBU message at time slot k with a fixed transmit power $P_U^{(k)}$ and a transmit energy cost $C_U^{(k)}$ if $x = 1$, and keeps silent otherwise. The system server can gather the sensing message sent from the OBU via RSU$_1$ or the UAV.

The jammer at a fixed location sends faked or replayed signals at time slot k to block RSU$_1$ with a jamming cost $C_J^{(k)}$ estimates the transmit power of the OBU and the UAV. In addition, the smart jammer applies smart radio devices to eavesdrop the control channel of the VANET and estimate the VANET transmission policy. According to the estimated VANET communication, the smart jammer changes its jamming power $y \in [0, P_J^M]$, where P_J^M is the maximum jamming power. For simplicity, we assume a constant noise power in the received signal denoted by σ and the jammer is too far away from the UAV and RSU$_2$ to block them.

5.2.2   Channel Model

The channel power gain vector of the system denoted by $\mathbf{h}^{(k)} = [h_{OR}^{(k)}, h_{OU}^{(k)}, h_{JR}^{(k)}, h_{JU}^{(k)}, h_{UR}^{(k)}]$ at time slot k consists of the channel power gain of the OBU-RSU$_1$ link $h_{OR}^{(k)}$, the OBU-UAV link $h_{OU}^{(k)}$, the jammer-RSU$_1$ link $h_{JR}^{(k)}$, the jammer-UAV link $h_{JU}^{(k)}$, and the UAV-RSU$_2$ link $h_{UR}^{(k)}$. Similar to [8], the channel gain is modeled as

$$h_i^{(k)} = \theta_0 \Delta_i^{(k)} \left(\frac{d_i^{(k)}}{d_0}\right)^{-\alpha_i} \tag{5.2}$$

Table 5.1 Summary of symbols and notations

$P^{(k)}$	Transmit power of the OBU at time slot k
$P_U^{(k)}$	Transmit power of the UAV at time slot k
P_J^M	Maximum jamming power of the smart jammer
V	Maximum OBU speed
$v^{(k)}$	Speed of the OBU at time slot k
σ	Received noise power
$\mathbf{h}^{(k)}$	Channel gain vector at time slot k
$\mathbf{d}^{(k)}$	Distance vector at time slot k
$P_e^{(k)}$	BER of the OBU message at time slot k
$\rho_O R^{(k)}$	SINR of the signals received by RSU$_2$ from the OBU at time slot k
$\rho_O U^{(k)}$	SINR of the signals received by UAV from the OBU at time slot k
$\rho_U R^{(k)}$	SINR of the signals received by RSU$_2$ from the UAV at time slot k
$x^{(k)}$	Relay action of the UAV at time slot k
$y^{(k)}$	Jamming power at time slot k
$\mathbf{A}$	Action set of the UAV
α	Learning rate of the UAV
δ	Learning discount factor of the UAV
$u_{U/J}^{(k)}$	Utility of the UAV/jammer at time slot k
$U_{U/J}^{(k)}$	Expected utility of the UAV/jammer at time slot k
$C_{U/J}^{(k)}$	Transmit cost of the UAV/jammer at time slot k

where θ_0 is the channel power gain at the reference distance d_0, and the channel time variation $\Delta_i^{(k)}$ depends on the Doppler shift due to the node mobility with $i \in \{OR, OU, JR, JU, UR\}$. The path-loss exponent α_i is set according to [17] and [18], e.g., $\alpha_i = 2$ for the OBU-UAV, jammer-UAV and UAV-RSU$_2$ links and $\alpha_i = 4$ otherwise. The path loss of the jammer-UAV radio link is assumed to be much higher than that of the jammer-RSU$_1$ link due to a longer distance. Similarly, the radio link between the OBU and the UAV has a higher path loss compared with that of the OBU-RSU$_1$ link. For ease of reference, some important notations are summarized in Table 5.1.

Due to the uncorrelated locations of RSU$_1$ and the UAV, $h_{OR}^{(k)}$ is independent with $h_{OU}^{(k)}$. The UAV transmission fails if the UAV is too far away from the OBU, and the UAV can cover the whole geographic area if it is high enough. For simplicity, the channel power gain is quantized into N levels with $h_i^{(k)} \in \{H_a\}_{1 \leq a \leq N}$, and is modeled as a Markov chain with N states. As shown in Fig. 5.2, the transition probability of the channel gain h_i from H_m to H_n during time slot k denoted by $p_{i,m,n}^{(k)}$ depends on the OBU speed given by

$$p_{i,m,n}^{(k)} = \Pr\left(h_i^{(k)} = H_n \mid h_i^{(k-1)} = H_m\right). \tag{5.3}$$

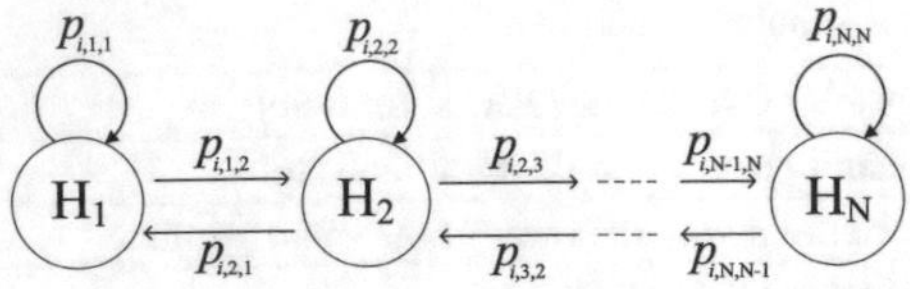

Fig. 5.2 Markov chain based channel model between the OBU and RSU_1 with N states, where $p_{i,m,n}$ is the probability that h_i changes from H_m to H_n in a time slot

5.2.3 Anti-jamming Transmission Game

The interactions between the UAV and the jammer are formulated as an anti-jamming relay game. In this game, the UAV decides whether or not to relay the OBU message to RSU_2, $x \in \mathbf{A}$, while the smart jammer chooses its jamming power $y \in [0, P_J^M]$. The UAV relay decision depends on the channel quality and the BER. The utility of the UAV at time slot k denoted by $u_U^{(k)}$ is based on the SINR of the signal received by the RSUs and the UAV and the transmit cost, which is given by

$$u_U^{(k)}(x, y) = x \max \left(\frac{P^{(k)} h_{OR}^{(k)}}{\sigma + y h_{JR}^{(k)}}, \min \left(\frac{P^{(k)} h_{OU}^{(k)}}{\sigma + y h_{JU}^{(k)}}, \frac{P_U^{(k)} h_{UR}^{(k)}}{\sigma} \right) \right)$$
$$- x C_U^{(k)} + \frac{(1-x) P^{(k)} h_{OR}^{(k)}}{\sigma + y h_{JR}^{(k)}}. \tag{5.4}$$

In (5.4), the first term in the max function represents the SINR of the signal sent by the OBU and received by RSU_1 at time slot k, and the second term represents the SINR of the weaker signal over the OBU-UAV link and the UAV-RSU_2 link at time slot k.

The utility of the jammer at time slot k, denoted by $u_J^{(k)}$, depends on the energy consumption of the UAV and the jamming cost, and is given by

$$u_J^{(k)}(x, y) = -x \max \left(\frac{P^{(k)} h_{OR}^{(k)}}{\sigma + y h_{JR}^{(k)}}, \min \left(\frac{P^{(k)} h_{OU}^{(k)}}{\sigma + y h_{JU}^{(k)}}, \frac{P_U^{(k)} h_{UR}^{(k)}}{\sigma} \right) \right)$$
$$+ x C_U^{(k)} + \frac{(x-1) P^{(k)} h_{OR}^{(k)}}{\sigma + y h_{JR}^{(k)}} - y C_J^{(k)}. \tag{5.5}$$

The time index k in the superscript is omitted unless necessary.

The NE of the anti-jamming game denoted by (x^*, y^*) corresponds to the best response strategy if the opponent follows the NE strategy. By definition, we have

$$u_U(x^*, y^*) \geq u_U(x, y^*), \quad \forall x \in \{0, 1\} \tag{5.6}$$

$$u_J(x^*, y^*) \geq u_J(x^*, y), \ \forall y \in \left[0, P_J^M\right]. \tag{5.7}$$

Theorem 5.1 *The anti-jamming transmission game in the UAV-aided VANET has an NE* $(0, 0)$, *if*

$$\sigma^2 > \min\left(\frac{Ph_1 h_{JR}}{C_J}, \min\left(\frac{(P_U h_5 - Ph_1)^2}{C_U^2},\right.\right.$$

$$\left.\left.\frac{P^2 (h_{OU} - h_{OR})^2}{C_U^2}\right)\right) \tag{5.8}$$

or

$$Ph_1 > \max(P_U h_5, Ph_2). \tag{5.9}$$

Proof By (5.4), if (5.8) holds, we have

$$u_U(1, 0) = \frac{P_U h_5}{\sigma} - C_U \leq \frac{Ph_1}{\sigma} = u_U(0, 0). \tag{5.10}$$

Thus, (5.6) holds for $(x^*, y^*) = (0, 0)$.
   By (5.5), let $\tilde{y} = \sqrt{Ph_1/(C_J h_3)} - \sigma/h_{JR} < 0$, we have

$$u_J(0, y) = -\frac{Ph_1}{\sigma + yh_3} - yC_J \tag{5.11}$$

$$\frac{\partial u_J(0, \tilde{y})}{\partial y} = \frac{Ph_1 h_{JR}}{(\sigma + \tilde{y} h_{JR})^2} - C_J = 0 \tag{5.12}$$

and $\partial^2 u_J / \partial y^2 \leq 0$, indicating that if (5.8) holds, $u_J(0, y)$ is concave with respect to (w.r.t.) y. Thus, $\forall y \in [0, P_J^M]$, we have

$$u_J(0, 0) = -\frac{Ph_1}{\sigma} \geq -\frac{Ph_1}{\sigma + yh_3} - yC_J = u_J(0, y). \tag{5.13}$$

Thus, (5.7) holds for $(x^*, y^*) = (0, 0)$, and thus $(0, 0)$ is an NE of the game. Similarly, we can prove that $(0, 0)$ is an NE of the game if (5.9) holds.

Note that the proof for the following theorems is similar to the proof of Theorem 5.1 and is omitted. Now we consider the BER of the OBU message in the VANET with quadrature phase-shift keying (QPSK). By (5.1), we have

$$P_e = \frac{1}{2} \min\left(\text{erfc}\left(\sqrt{\frac{\rho_1}{2}}\right), \text{erfc}\left(\sqrt{\frac{\min(\rho_2, \rho_3)}{2}}\right)\right). \tag{5.14}$$

Corollary 5.1 *If (5.8) holds, the BER of the OBU message in the UAV-aided VANET with QPSK at the NE of the game is given by*

$$P_e = \frac{1}{2}\text{erfc}\left(\sqrt{\frac{\min(Ph_{OU}, P_U h_5)}{2\sigma}}\right). \tag{5.15}$$

Corollary 5.2 *If (5.9) holds, the BER of the OBU message in the UAV-aided VANET with QPSK at the NE of the game is given by*

$$P_e = \frac{1}{2}\text{erfc}\left(\sqrt{\frac{Ph_{OR}}{2\sigma}}\right). \tag{5.16}$$

Remark Both the jammer and the UAV should keep silent to reduce the power consumption if the jamming cost and the transmit cost are high as shown in (5.8) or the link between the OBU and RSU$_1$ is in a good condition as shown in (5.9). In the case, the BER of the OBU message decreases with the transmit power of the OBU as shown in (5.15) and (5.16).

Theorem 5.2 *The anti-jamming transmission game in the UAV-aided VANET has an NE $(1, 0)$, if*

$$C_U\sigma + Ph_{OR} \leq P_U h_5 < \frac{Ph_{OU}\sigma}{\sigma + P_J^M h_4} \tag{5.17}$$

or

$$\frac{C_U\sigma}{h_{OU} - h_{OR}} \leq P < \min\left(\frac{C_J\sigma^2}{h_{OR}h_{JU}}, \frac{P_U h_5}{h_{OU}}\right) \tag{5.18}$$

$$h_{JU} < h_{JR}. \tag{5.19}$$

Corollary 5.3 *If (5.17) holds, the BER of the OBU message in the UAV-aided VANET with QPSK at the NE of the game is given by*

$$P_e = \frac{1}{2}\text{erfc}\left(\sqrt{\frac{P_U h_5}{2\sigma}}\right). \tag{5.20}$$

Corollary 5.4 *If (5.18) and (5.19) hold, the BER of the OBU message in the UAV-aided VANET with QPSK at the NE of the game is given by*

$$P_e = \frac{1}{2}\text{erfc}\left(\sqrt{\frac{Ph_{OU}}{2\sigma}}\right). \tag{5.21}$$

Remark The UAV relays for the OBU to improve its performance if the transmit cost is low and the link between the OBU and the UAV is in a good condition as shown in (5.17). The jammer keeps silent if the jamming cost is high as shown in (5.18).

Theorem 5.3 *The anti-jamming transmission game in the UAV-aided VANET has an NE* $\left(0, \sqrt{Ph_1/(C_J h_3)} - \sigma/h_{JR}\right)$, *if*

$$
\frac{Ph_{OR}h_{JR}}{(\sigma + P_J^M h_3)^2} \le C_J \le \min\left(\frac{Ph_1 h_{JR}}{\sigma^2}, \right.
$$

$$
\left. \max\left(\frac{P_U^2 h_{UR}^2 h_{JR}}{Ph_{OR}\sigma^2}, \frac{Ph_{JR}(h_{OU}h_{JR} - h_{OR}h_{JU})^2}{h_{OR}\sigma^2(h_{JR} - h_{JU})^2}\right)\right) \tag{5.22}
$$

$$
C_U \ge \min\left(\frac{P_U h_5}{\sigma} - \frac{Ph_{OR}}{\sigma + P_J^M h_3}, \frac{P(h_{OU} - h_{OR})}{\sigma}\right) \tag{5.23}
$$

or

$$
Ph_1 C_J \sigma^2 \ge \max\left(P_U^2 h_{UR}^2 h_{JR}, \right.
$$

$$
\left. \frac{P^2 h_{OU}^2 h_{JR}^3 \sigma^2 C_J}{\left(\sigma(h_{JR} - h_{JU})\sqrt{C_J} + h_{JU}\sqrt{Ph_{OR}h_{JR}}\right)^2}\right). \tag{5.24}
$$

Theorem 5.4 *The anti-jamming transmission game in the UAV-aided VANET has an NE* $\left(1, \sqrt{Ph_2/(C_J h_4)} - \sigma/h_{JU}\right)$, *if*

$$
C_U \le \frac{Ph_{OU}}{\sigma + P_J^M h_4} - \frac{Ph_{OR}}{\sigma + P_J^M h_3} \tag{5.25}
$$

$$
\max\left(\frac{Ph_{OR}h_{JU}}{(\sigma + P_J^M h_4)^2}, \frac{Ph_{JU}(h_{OR}h_{JU} - h_{OU}h_{JR})^2}{h_{OU}\sigma^2(h_{JU} - h_{JR})^2}\right)
$$

$$
\le C_J \le \min\left(\frac{Ph_{OR}h_{JU}}{\sigma^2}, \frac{P_U^2 h_{UR}^2 h_{JU}}{Ph_{OU}\sigma^2}\right). \tag{5.26}
$$

Theorem 5.5 *The anti-jamming transmission game in the UAV-aided VANET has an NE* $\left(0, P_J^M\right)$, *if*

$$
C_J\left(\sigma + P_J^M h_3\right)^2 < Ph_{OR}h_{JR} \tag{5.27}
$$

$$
C_U \ge \min\left(\frac{P_U h_5}{\sigma} - \frac{Ph_{OR}}{\sigma + P_J^M h_3}, \right.
$$

$$\left. \frac{Ph_{OU}}{\sigma + P_J^M h_4} - \frac{Ph_{OR}}{\sigma + P_J^M h_3} \right) \tag{5.28}$$

or

$$Ph_{OR}\sigma > \max\left(P_U h_5(\sigma + P_J^M h_3), Ph_{OU}\sigma\right). \tag{5.29}$$

Theorem 5.6 *The anti-jamming transmission game in the UAV-aided VANET has an NE* $(1, P_J^M)$, *if*

$$C_U \leq \frac{Ph_{OU}}{\sigma + P_J^M h_4} - \frac{Ph_{OR}}{\sigma + P_J^M h_3} \tag{5.30}$$

$$\frac{C_J(\sigma + P_J^M h_4)^2}{h_{OR}h_{JU}} < P \leq \frac{P_U h_5}{h_{OU}} \tag{5.31}$$

$$h_{OU}\sigma > h_{OR}(\sigma + P_J^M h_4). \tag{5.32}$$

Numerical results with $P = 10$, $P_U = 2$, $\sigma = 0.1$, $h_{OR} = 0.2$, $h_{JR} = 0.4$, $h_{JU} = 0.2$ and $h_{UR} = 0.3$ in Fig. 5.3 show that the BER of the OBU message at the NE of the game decreases with the jamming cost C_J. For instance, the BER decreases from 0.48% to 0.18% and the utility of the UAV increases by 38%, as C_J changes from 0.5 to 0.8, because the smart jammer is less motivated to attack the VANET under a higher jamming cost. The utility of the UAV increases with the channel gain h_{OU}, and the utility of the jammer decreases with it. For example, the utility of the UAV increases by 47.5% as the channel gain h_{OU} changes from 0.45 to 0.8.

5.2.4   Anti-jamming Transmission Stochastic Game

In the anti-jamming relay stochastic game, the UAV decides whether or not relay the OBU message, $x \in \mathbf{A}$, and the smart jammer chooses its jamming power $y \in [0, P_J^M]$ under the time-variant random channel power gains, which can be modeled as a Markov chain with N states. As shown in Fig. 5.2, the transition probability of h_i at time slot k in the Markov chain based channel model is denoted by $\mathbf{p}_i^{(k)} = [p_{i,n}^{(k)}]_{1 \leq n \leq N}$, where $p_{i,n}^{(k)}(h_i^{(k-1)}) = \Pr(h_i^{(k)} = H_n \mid h_i^{(k-1)})$ with $i \in \{OR, OU, JR, JU, UR\}$.

According to the channel model in Fig. 5.2, the expected utility of the UAV in the stochastic game at time slot k over the $N \times N$ realizations of the channel power gains, $h_{OR}^{(k)}$ and $h_{OU}^{(k)}$, is denoted by $U_U^{(k)}$, and given by (5.4) as

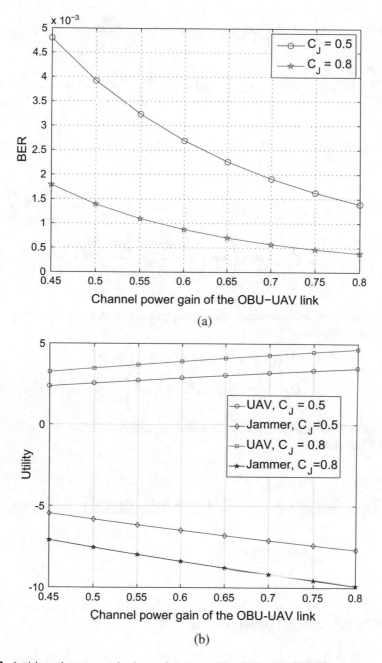

Fig. 5.3 Anti-jamming communication performance of the UAV-aided VANET in the game at the NE, with $P = 10$, $P_U = 2$, $C_U = 1$, $\sigma = 0.1$ and $\mathbf{h} = [0.2, h_{OU}, 0.4, 0.2, 0.3]$. (**a**) BER of the OBU message. (**b**) Utility

$$U_U^{(k)}(x, y) = \mathbb{E}_{h_{OR}^{(k)}, h_{OU}^{(k)}} \left[u_U^{(k)}(x, y) \right] = \sum_{m=1}^{N} \frac{(1 - x) p_{OR,m}^{(k)} P^{(k)} H_m}{\sigma + y h_{JR}^{(k)}} - x C_U^{(k)}$$

$$+ \sum_{m=1}^{N} \sum_{n=1}^{N} p_{OR,m}^{(k)} p_{OU,n}^{(k)} x \max \left(\min \left(\frac{P^{(k)} H_n}{\sigma + y h_{JU}^{(k)}}, \frac{P_U^{(k)} h_{UR}^{(k)}}{\sigma} \right), \frac{P^{(k)} H_m}{\sigma + y h_{JR}^{(k)}} \right).$$
(5.33)

Similarly, the expected utility of the jammer at time slot k in the stochastic game is denoted by $U_J^{(k)}$, and given by (5.5) as

$$U_J^{(k)}(x, y) = \mathbb{E}_{h_{OR}^{(k)}, h_{OU}^{(k)}} \left[u_J^{(k)}(x, y) \right] = x C_U^{(k)} - y C_J^{(k)} + \sum_{m=1}^{N} \frac{(x - 1) p_{OR,m}^{(k)} P^{(k)} H_m}{\sigma + y h_{JR}^{(k)}}$$

$$- \sum_{m=1}^{N} \sum_{n=1}^{N} p_{OR,m}^{(k)} p_{OU,n}^{(k)} x \max \left(\min \left(\frac{P^{(k)} H_n}{\sigma + y h_{JU}^{(k)}}, \frac{P_U^{(k)} h_{UR}^{(k)}}{\sigma} \right), \frac{P^{(k)} H_m}{\sigma + y h_{JR}^{(k)}} \right).$$
(5.34)

The NE of the anti-jamming stochastic game denoted by (x^*, y^*) is given by definition as follows:

$$U_U(x^*, y^*) \geq U_U(x, y^*), \ \forall x \in \{0, 1\}$$
(5.35)

$$U_J(x^*, y^*) \geq U_J(x^*, y), \ \forall y \in \left[0, P_J^M \right].$$
(5.36)

Theorem 5.7 *The anti-jamming transmission stochastic game in the UAV-aided VANET has an NE $(0, 0)$, if*

$$\sigma \geq \max$$

$$\times \left(\frac{\sum_{m=1}^{N} \sum_{n=1}^{N} P_{OR,m} P_{OU,n} \max \left(P H_m, \min \left(P H_n, P_U h_{UR} \right) \right) - \sum_{m=1}^{N} P_{OR,m} P H_m}{C_U}, \right.$$

$$\left. \sqrt{\frac{\sum_{m=1}^{N} P_{OR,m} P H_m h_{JR}}{C_J}} \right).$$
(5.37)

Proof By (5.33), if (5.37) holds, we have

$$U_U(1, 0) = \sum_{m=1}^{N} \sum_{n=1}^{N} P_{OR,m} P_{OU,n} \max \left(\frac{P H_m}{\sigma}, \min \left(\frac{P H_n}{\sigma}, \frac{P_U h_{UR}}{\sigma} \right) \right) - C_U$$

$$\leq \sum_{m=1}^{N} \frac{P_{OR,m} P H_m}{\sigma} = U_U(0, 0).$$
(5.38)

Thus, (5.35) holds for $(x^*, y^*) = (0, 0)$.

By (5.34), if (5.37) holds, we have

$$\frac{\partial U_J(0, y)}{\partial y} = \sum_{m=1}^{N} \frac{p_{OR,m} P H_m h_{JR}}{(\sigma + y h_{JR})^2} - C_J < 0 \tag{5.39}$$

indicating that $U_J(0, y)$ decreases with y, and $\forall y \in [0, P_J^M]$ we have

$$U_J(0, 0) = -\sum_{m=1}^{N} \frac{p_{OR,m} P H_m}{\sigma} \tag{5.40}$$

$$\geq -\sum_{m=1}^{N} \frac{p_{OR,m} P H_m}{\sigma + y h_{JR}} - y C_J = U_J(0, y).$$

Similarly, we can prove that (5.36) also holds for $(x^*, y^*) = (0, 0)$, and thus $(0, 0)$ is an NE of the game.

Remark Both the jammer and the UAV should keep silent to reduce the power consumption if the jamming cost and the UAV transmit cost are high as shown in (5.37).

Theorem 5.8 *The anti-jamming transmission stochastic game in the UAV-aided VANET has an NE* $(1, 0)$*, if*

$$\sqrt{\frac{\sum_{m=1}^{N} p_{OR,m} P H_m h_{JR}}{C_J}} < \sigma$$

$$< \frac{\sum_{m=1}^{N} \sum_{n=1}^{N} p_{OR,m} p_{OU,n} \max \left(P H_m, \min \left(P H_n, P_U h_{UR} \right) \right)}{C_U}$$

$$- \frac{\sum_{m=1}^{N} p_{OR,m} P H_m}{C_U}. \tag{5.41}$$

Proof The proof is similar to that of Theorem 5.7.

Remark The UAV relays for the OBU to improve its performance if the transmit cost is low and the link between the OBU and the UAV is in a good condition, and the jammer keeps silent if the jamming cost is high as shown in (5.41).

Numerical results with $P = 10 \, \text{mW}$, $P_U = 2 \, \text{mW}$, $h = [h_{OR}, h_{OU}, 0.4, 0.2, 0.5]$, $C_U = 1$ and $C_J = 0.8$ as shown in Fig. 5.4 show that the BER of the OBU message at the NE of the stochastic game increases with the transition probability of h_{OR} and that of h_{OU}. For instance, the BER increases by 72% and the utility of the UAV decreases by 13.3% as $p_{OU,1,2}$ changes from 0.5 to 0.8, because the UAV has more trouble to gain the optimal strategies with the unstable channel conditions.

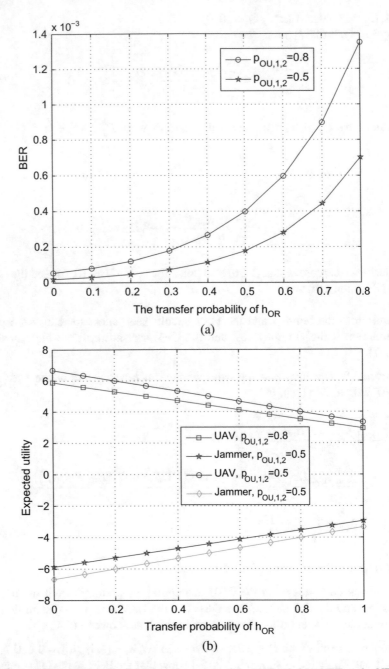

Fig. 5.4 Anti-jamming communication performance of the UAV-aided VANET at the NE, with $P = 10$, $P_U = 2$, $\mathbf{h} = [h_{OR}, h_{OU}, 0.4, 0.2, 0.5]$, $C_U = 1$ and $C_J = 0.8$. (**a**) BER of the OBU message. (**b**) Expected utility

5.3 Reinforcement Learning-Based UAV Relay Scheme for VANETs

Reinforcement learning techniques have been widely applied to improve security in wireless networks [19–21]. The non-cooperative power control algorithm as presented in [21] in the repeated game can improve the throughput of wireless ad hoc networks. The prospect theory based dynamic game as formulated in [19] shows the impact of the subjectivity of end-users and jammers on the throughput of cognitive radio networks with Q-learning algorithm. The deep Q-network algorithm as proposed in [20] uses both frequency and spatial diverting to improve the SINR of the signals and the utility of the secondary user in cognitive radio networks.

The UAV relay process in the dynamic game can be viewed as an MDP, and thus reinforcement learning (RL) such as Q-learning can achieve the optimal strategy via trials-and-errors if the game is long enough [19, 20]. In this section, the repeated interactions between the UAV and the smart jammer in the VANET can be formulated as a dynamic game, in which the jammer determines its jamming power based on the previous VANET transmission, and the UAV chooses its relay strategy based on the system state, which consists of the radio channel state and the BER of the OBU message observed in last time slot. The next system state observed by the UAV is independent of the previous states and actions, for a given system state and UAV relay strategy in the current time slot. Therefore, the UAV relay process in the dynamic game can be viewed as an MDP and the UAV can apply reinforcement learning techniques such as Q-learning to derive its optimal strategy via trials without the knowledge of jamming model.

In the dynamic game, the UAV decides whether or not to relay the OBU message based on the system state at time slot k denoted by $\mathbf{s}^{(k)}$ that consists of the link quality between the UAV and the OBU, that between RSU$_1$ and the OBU, the SINR between RSU$_2$ and the UAV, and the BER of the OBU message at the previous time slot, i.e., $\mathbf{s}^{(k)} = [\rho_{OU}^{(k-1)}, \rho_{OR}^{(k-1)}, \rho_{UR}^{(k-1)}, P_e^{(k-1)}]$.

The learning rate denoted by $\alpha \in (0, 1]$ shows the weight of the current experience, and the discount factor $\delta \in [0, 1]$ corresponds to the uncertainty on the future utility. The Q-function of the action x at state $\mathbf{s}$ is denoted by $Q(\mathbf{s}, x)$ and is updated in each time slot according to iterative Bellman equation as follows:

$$Q(\mathbf{s}, x) \leftarrow (1 - \alpha)Q(\mathbf{s}, x) + \alpha \left(u_U(\mathbf{s}, x) + \delta V(\mathbf{s}')\right), \qquad (5.42)$$

where $\mathbf{s}'$ is the next state if the UAV chooses x at state $\mathbf{s}$, and the value function $V(\mathbf{s})$ maximizes $Q(\mathbf{s}, x)$ over the UAV action set given by

$$V(\mathbf{s}) \leftarrow \max_{x \in \{0, 1\}} Q(\mathbf{s}, x). \qquad (5.43)$$

The UAV then selects whether or not to relay the OBU message $x \in \mathbf{A}$ according to the ϵ-greedy strategy, i.e.,

$$
\Pr(x = x^*) = \begin{cases} 1 - \epsilon, & x^* = \arg\max_{\mathbf{A}} Q(\mathbf{s}, x) \\ \frac{\epsilon}{|A|-1}, & \text{o.w.} \end{cases} \tag{5.44}
$$

5.4   Hotbooting PHC-Based UAV Relay Algorithm for VANETs

As a model-free RL technique for the mixed-strategy game, the policy hill climbing (PHC) algorithm can achieve the optimal policy without knowing the jamming model and the VANET model [22]. Compared with Q-learning, the PHC-based relay strategy that provides more randomness in the decision can fool the jammers with uncertainty. Therefore, we propose the PHC-based UAV relay strategy to improve the anti-jamming performance of the VANET communication. In this scheme, the jammer cannot predict the optimal relay policy and thus cannot attack the UAV accordingly.

Based on the PHC-based relay strategy, the UAV selects whether or not to relay the OBU message $x \in \mathbf{A}$ according to the mixed strategy. The mixed-strategy table in the PHC-based relay denoted by $\pi(\mathbf{s}, x)$ is updated by increasing the probability corresponding to the highest valued action by $\beta \in (0, 1]$, and decreasing other probabilities by $-\beta/(|\mathbf{A}| - 1)$, i.e.,

$$
\pi(\mathbf{s}, x) \leftarrow \pi(\mathbf{s}, x) + \begin{cases} \beta, & x = \max_{\hat{x} \in \{0,1\}} Q(\mathbf{s}, \hat{x}) \\ \frac{-\beta}{|A|-1}, & \text{o.w.} \end{cases} \tag{5.45}
$$

The UAV then selects whether or not to relay the OBU message $x \in \mathbf{A}$ according to the mixed strategy $\pi(\mathbf{s}, x)$, i.e.,

$$
\Pr\left(x = x^*\right) = \pi\left(\mathbf{s}, x^*\right), \quad x^* \in \mathbf{A}. \tag{5.46}
$$

As a type of transfer learning [23], the hotbooting technique is applied to accelerate the learning process and thus improve the VANET communication performance against jamming. More specifically, the hotbooting technique initializes the value of the quality function or Q-function for each action-state pair with the experiences in similar scenarios to avoid the random exploration of the standard PHC that initializes the Q-function with an all-zero matrix. A hotbooting technique as shown in Algorithm 5.1 exploits the experiences from large-scale UAV-aided VANET experiments to initialize the Q-value and the mixed-strategy π for each action-state pair to reduce the random exploration at the beginning of the repeated game and accelerate the learning process of the relay. More specifically, we consider ξ anti-

---

Algorithm 5.1 Hotbooting process in the UAV relay

1: Initialize α, β, δ, ξ, $s^{(0)}$, and $\mathbf{A}$
2: $\mathbb{E} = \emptyset$, $\mathbf{Q} = 0$, $\mathbf{V} = 0$, $\pi = \frac{1}{A}$
3: **for** $n = 1, 2, 3, \ldots, \xi$ **do**
4:     **for** $k = 1, 2, \ldots, K$ **do**
5:         Choose $x^{(k)} \in \mathbf{A}$ via (5.46)
6:         **if** $x^{(k)} = 1$ **then**
7:             Relay the OBU message to RSU$_2$ with a fixed transmit power $P_U^{(k)}$
8:         **end if**
9:         Receive the SINR ρ_{OU}, ρ_{OR}, ρ_{UR}, and the BER P_e from server
10:        Obtain utility $u_U^{(k)}$
11:        Calculate $Q^* \left(s^{(k)}, x^{(k)} \right)$ via (5.42) and (5.43)
12:        Calculate $\pi^* \left(s^{(k)}, x^{(k)} \right)$ via (5.45)
13:        $s^{(k+1)} = \left[\rho_{OU}^{(k)}, \rho_{OR}^{(k)}, \rho_{UR}^{(k)}, P_e^{(k)} \right]$
14:     **end for**
15: **end for**
16: Output $\mathbf{Q}^*$ and π^*

Algorithm 5.2 Hotbooting PHC-based UAV relay strategy

1: Call Algorithm 5.1
2: Initialize α, β, δ, $\mathbf{A}$, and $s^{(0)}$
3: $\mathbf{Q} = \mathbf{Q}^*$, $\mathbf{V} = 0$, $\pi = \pi^*$
4: **for** $k = 1, 2, \ldots$ **do**
5:     Choose $x^{(k)} \in \mathbf{A}$ via (5.46)
6:     **if** $x^{(k)} = 1$ **then**
7:         Relay the OBU message to RSU$_2$ with a fixed transmit power $P_U^{(k)}$
8:     **end if**
9:     Collect the SINR ρ_{OU}, ρ_{OR}, ρ_{UR}, and the BER P_e from server
10:    Obtain utility $u_U^{(k)}$
11:    Update $Q \left(s^{(k)}, x^{(k)} \right)$ via (5.42)
12:    Update $V \left(s^{(k)} \right)$ via (5.43)
13:    Update $\pi \left(s^{(k)}, x^{(k)} \right)$ via (5.45)
14:    $s^{(k+1)} = \left[\rho_{OU}^{(k)}, \rho_{OR}^{(k)}, \rho_{UR}^{(k)}, P_e^{(k)} \right]$
15: **end for**

---

jamming VANET transmission experiments in similar scenarios before the game. Each experiment that lasts K time slots, during which the UAV observes the current state, i.e., the anti-jamming transmission performance such as the BER of the OBU message in the VANET and then chooses the relay strategy with the mixed-strategy $\pi (s, x)$ according to (5.42), (5.43), (5.44), and (5.45).

The initial Q-values $\mathbf{Q}^*$ and the mixed-strategy table π^* as the output of Algorithm 5.1 via ξ experiments are used to initialize the Q-values and mixed-strategy table in Algorithm 5.2, with $\mathbf{Q} = \mathbf{Q}^*$ and $\pi = \pi^*$. As shown in Algorithm 5.2, the UAV observes the current state $s^{(k)}$ consisting of the channel quality of the OBU-RSU$_1$ link, the OBU-UAV link and the UAV-RSU$_2$ link and the BER of the OBU message at the previous time slot from the server. The relay decision $x^{(k)}$ is chosen according to (5.46). The utility of the UAV $u_U^{(k)}$ is evaluated and both the Q-function and mixed-strategy π are updated via (5.42), (5.43), (5.44), and (5.45) in each time slot. In this way, the UAV learns the jamming strategy in the anti-jamming transmission dynamic game and achieves the optimal relay strategy to improve the long-term anti-jamming transmission performance.

5.5   Performance Evaluation

Simulations have been performed to evaluate the performance of the proposed UAV relay strategy in the dynamic game against smart jamming. In the simulations, both the UAV and the jammer were stationary and observed the ongoing VANET communication. The jammer applied greedy strategy to choose the jamming power regarding the expected reward u_J in (5.5) during the VANET communication in each time slot. The radio link between the jammer and RSU$_2$ was much worse than that between the jammer and RSU$_1$ due to a longer distance, which leads to a lower channel gain. Unless specified otherwise, we set the transmit power of the OBU $P = 10$ mW, the transmit power of the UAV $P_U = 2$ mW, the transmit energy cost $C_U = 1$, the jamming cost $C_J = 0.5$, $\sigma = 0.1$, $h_{JR} = 0.4$ and $h_{JU} = 0.2$. More specifically, the transmit power of the UAV was 5 times higher than that of the OBU, with a unit relay energy cost. The jammer had a lower channel gain to the UAV than RSU$_1$ due to a longer distance, i.e., $h_{JR} = 0.4 > h_{JU} = 0.2$. Similar to the channel model in [24], the Doppler shift was considered in the OBU-RSU$_1$ and OBU-UAV channel models. The channel gain transition probability linearly increases with the moving speed of the OBU, and is given by

$$
p_{i,m,n}^{(k)} = \begin{cases} \frac{\varphi}{V} v^{(k)}, & \text{if } (m,n) = (1,2) \text{ or } (N, N-1) \\ 1 - \frac{\varphi}{V} v^{(k)}, & \text{if } 1 \leq m = n \leq N \\ \frac{\varphi}{2V} v^{(k)}, & \text{if } 2 \leq m \leq N-1 \text{ and } n = m \pm 1 \\ 0, & \text{otherwise,} \end{cases} \tag{5.47}
$$

where $i \in \{OR, OU, JR, JU, UR\}$ and φ denotes the impact of the environmental changes.

As shown in Fig. 5.5, the BER of the OBU message decreases with time, e.g., from 1.41% at the beginning of the game to 0.3% after 1500 time slots, which is about 65.7% lower than that of the Q-learning based strategy. This is because the PHC-based relay strategy increases the randomness of the UAV compared with Q-learning. The utility of the UAV increases by about 52.2% after 1500 time slots, which is about 42% higher than Q-learning. Our proposed scheme takes about 500 time slots (about 0.45 s) to converge to the optimal policy under static network environment and keeps using it unless the network state or the attack policy changes. The convergent speed of the hotbooting PHC-based relay is 56.3% faster than the standard PHC, due to the hotbooting technique formulates the emulated experiences.

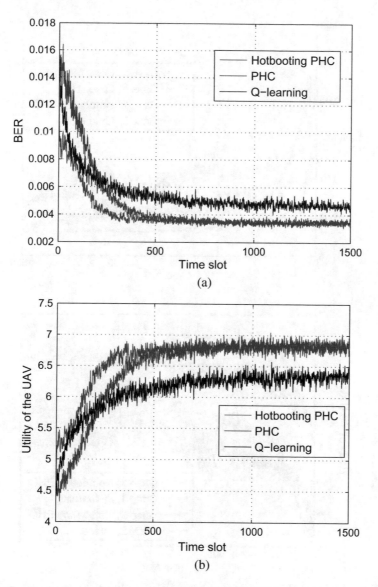

Fig. 5.5 Communication performance of the UAV-aided VANET in the dynamic game, with $C_U = 1$ and $C_J = 0.5$. (**a**) BER of the OBU message. (**b**) Utility of the UAV

The anti-jamming transmission performance of the dynamic game is shown in Fig. 5.6, in which a smart jammer changed its jamming power according to the expected learning results of the UAV relay every 500 time slots in Case 1. The channel conditions changed randomly every 100 time slots in Case 2. The proposed strategy is more robust against the smart jammer, e.g., our proposed scheme reduces

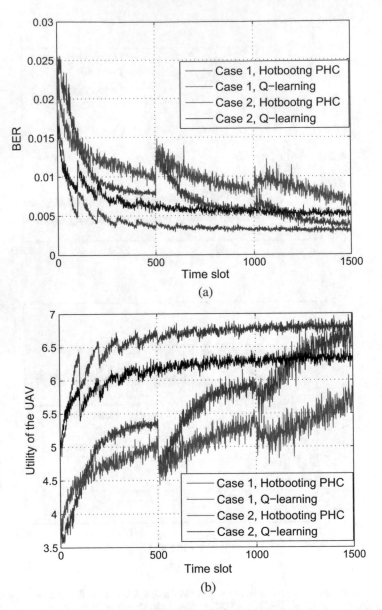

Fig. 5.6 Communication performance of the UAV-aided VANET in the dynamic game against the smart jammer that changes the jamming policy every 500 time slots in Case 1, with the channel conditions changing randomly every 100 time slots in Case 2. (**a**) BER of the OBU message. (**b**) Utility of the UAV

the BER of the OBU message in Case 1 by 40.9% and increases the utility of the UAV by 18.3% compared with Q-learning at time slot 1000. In Case 2, the BER of the OBU message of the hotbooting PHC-based relay is 37.5% lower than Q-learning at time slot 100. Our proposed scheme applies the hotbooting technique to accelerate the learning process in the jamming resistance, yielding a faster learning speed compared with the radio channel variations.

The BERs of the OBU messages averaged over 1500 time slots (approximately 1.36 s) are presented in Fig. 5.7, showing that the BER increases with the UAV-RSU$_2$ distance and decreases with the jammer-UAV distance. For example, the BER decreases from 0.94% to 0.23% as d_{UR} changes from 60 to 50 m, which is 30.4% lower than the Q-learning based strategy. The average utility of the UAV decreases with the UAV-RSU$_2$ distance and increases with the jammer-UAV distance, e.g., the average utility increases by 19% as d_{JU} changes from 50 to 60 m. That is because the UAV has a better radio link with OBU and RSU$_2$ due to the longer distance from the jammer.

The performance of the dynamic relay game is shown in Fig. 5.8. The average BER of the OBU messages increases with the OBU speed, e.g., from 0.3% to 0.65% as v changes from 4 to 18 m/s with $C_U = 0.5$ and $C_J = 0.1$, which is about 63% lower than Q-learning. That is because the high mobility of the OBU increases the transition probabilities of h_{OR} and h_{OU} and leads to an unstable channel condition. The average utility of the UAV decreases by about 24% as v changes from 4 to 18 m/s, which is about 21% lower than Q-learning. The average BER with $C_U = 1$ and $C_J = 0.5$ is higher than that with $C_U = 0.5$ and $C_J = 0.1$, because the UAV has less motivation to relay the OBU message with a higher transmission cost.

5.6 Summary

In this chapter, we have formulated a UAV-aided VANET relay stochastic game and derived the NEs of the game to disclose the impacts of the UAV transmit cost and the radio channel condition on communication performance of the VANET against smart jamming. A hotbooting PHC-based anti-jamming relay strategy has been proposed for a UAV to achieve the optimal relay policy without knowing the VANET model and the jamming model. Simulation results have shown that the proposed relay strategy can efficiently improve the anti-jamming transmission performance of the VANET. For example, this scheme reduces the BER of the OBU message in the VANET by 65.7% and increases the utility of the UAV by 42% compared with the Q-learning based relay strategy.

We will carry out a more in-depth game theoretic study of the anti-jamming UAV relay in VANETs in the future, which incorporates more jamming models, including a smart jammer that can choose both the jamming strength and the mobility. A backup anti-jamming transmission mechanism need to be provided

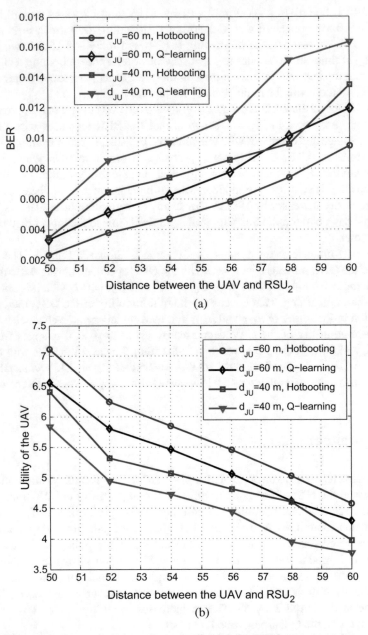

Fig. 5.7 Communication performance of the UAV-aided VANET in the dynamic game with $d_{JU} = 60$ m in Case 1, and $d_{JU} = 50$ m in Case 2. (**a**) Average BER of the OBU messages. (**b**) Average utility of the UAV

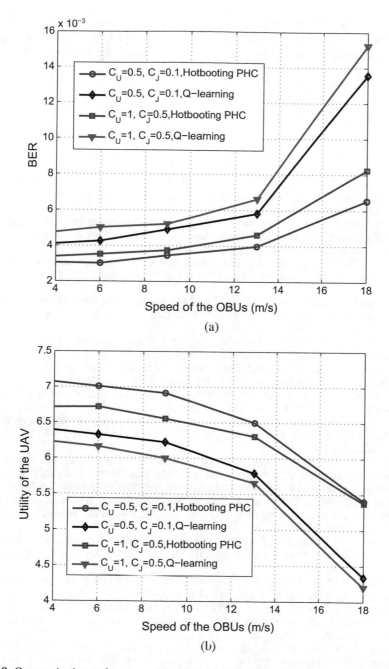

Fig. 5.8 Communication performance of the UAV-aided VANET in the dynamic game with $C_U = 0.5$ and $C_J = 0.1$ in Case 1, and $C_U = 1$ and $C_J = 0.5$ in Case 2. (**a**) Average BER of the OBU messages. (**b**) Average utility of the UAV

to protect VANETs at the beginning stage of the learning process. In addition, we should develop new RL techniques with low computation and communication overhead to improve the anti-jamming communication performance of UAV-aided VANETs.

References

1. H. Hartenstein and L. Laberteaux, "A tutorial survey on vehicular ad hoc networks," *IEEE Commun. Mag.*, vol. 46, no. 6, pp. 164–171, Jun. 2008.
2. I. K. Azogu, M. T. Ferreira, J. A. Larcom, and H. Liu, "A new anti-jamming strategy for VANET metrics-directed security defense," in *Proc. IEEE Global Commun. Conf.*, pp. 1344–1349, Atlanta, GA, Dec. 2013.
3. H. Sedjelmaci, S. M. Senouci, and N. Ansari, "Intrusion detection and ejection framework against lethal attacks in UAV-aided networks: A bayesian game-theoretic methodology," *IEEE Trans. Intell. Transportation Syst.*, pp. 1–11, Aug. 2016.
4. Y. Zhou, N. Cheng, N. Lu, and X. Shen, "Multi-UAV-aided networks: Aerial-ground cooperative vehicular networking architecture," *IEEE Veh. Technol. Mag.*, vol. 10, no. 4, pp. 36–44, Dec. 2015.
5. P. Zhan, K. Yu, and A. L. Swindlehurst, "Wireless relay communications with unmanned aerial vehicles: Performance and optimization," *IEEE Trans. Aerospace and Electronic Systems*, vol. 47, no. 3, pp. 2068–2085, Jul. 2011.
6. C. Dixon and E. W. Frew, "Optimizing cascaded chains of unmanned aircraft acting as communication relays," *IEEE J. Sel. Areas Commun.*, vol. 30, no. 5, pp. 883–898, Jun. 2012.
7. Y. Zeng, R. Zhang, and T. J. Lim, "Throughput maximization for UAV-enabled mobile relaying systems," *IEEE Trans. Commun.*, vol. 64, no. 12, pp. 4983–4996, Dec. 2016.
8. Q. Wang, Z. Chen, W. Mei, and J. Fang, "Improving physical layer security using UAV-enabled mobile relaying," *IEEE Wireless Comm. Letters*, Mar. 2017.
9. J. Kosmerl and A. Vilhar, "Base stations placement optimization in wireless networks for emergency communications," in *Proc. IEEE Int. Conf. Commun.*, pp. 200–205, Sydney, Australia, Jun. 2012.
10. G. Tuna, T. V. Mumcu, K. Gulez, V. C. Gungor and H. Erturk, "Unmanned aerial vehicle-aided wireless sensor network deployment system for post-disaster monitoring," in *Emerging Intelligent Computing Technology and Applications*, vol. 304, pp. 298–305, Jul. 2012.
11. T. A. Johansen, A. Zolich, and T. Hansen, "Unmanned aerial vehicle as communication relay for autonomous underwater vehicle-Field tests," in *Proc. IEEE Global Commun. Conf.*, pp. 1469–1474, Austin, TX, Dec. 2014.
12. J. Ueyama, H. Freitas, B. S. Faiçal, et al., "Exploiting the use of unmanned aerial vehicles to provide resilience in wireless sensor networks," *IEEE Commun. Mag.*, vol. 52, no. 12, pp. 81–87, Dec. 2014.
13. M. Dong, K. Ota, M. Lin, and et al., "UAV-assisted data gathering in wireless sensor networks," *J. Supercomput.*, vol. 70, no. 3, pp. 1142–1155, Dec. 2014.
14. M. Mozaffari, W. Saad, M. Bennis, and M. Debbah, "Unmanned aerial vehicle with underlaid device-to-device communications," *IEEE Trans. Wireless Commun.*, vol. 15, no. 6, pp. 3949–3963, Jun. 2016.
15. A. Sanjab, W. Saad, and T. Başar, "Prospect theory for enhanced cyber-physical security of drone delivery systems: A network interdiction game," *Proc. IEEE Int. Conf. Commun. (ICC)*, Paris, France, May 2017.
16. L. Xiao, *Anti-jamming transmissions in cognitive radio networks*. Springer, 2015.
17. V. Erceg, L. J. Greenstein, S. Y. Tjandra, and et al., "An empirically based path loss model for wireless channels in suburban environments," *IEEE J. Sel. Areas Commun.*, vol. 17, no. 7, pp. 1205–1211, Jul. 1999.

18. R. Palat, A. Annamalau, and J. Reed, "Cooperative relaying for ad-hoc ground networks using swarm UAVs," in *Proc. IEEE Military Commun. Conf.*, pp. 1588–1594, Atlantic, NJ, Oct. 2005.

19. L. Xiao, J. Liu, Q. Li, N. B. Mandayam, and H. V. Poor, "User-centric view of jamming games in cognitive radio networks," *IEEE Trans. Inf. Forensics and Security*, vol. 10, no. 12, pp. 2578–2590, Dec. 2015.

20. G. Han, L. Xiao, and H. V. Poor, "Two-dimensional anti-jamming communication based on deep reinforcement learning," in *Proc. IEEE Int. Conf. Acoustics, Speech and Signal Processing (ICASSP)*, pp. 1–5, New Orleans, LA, Mar. 2017.

21. C. Long, Q. Zhang, B. Li, H. Yang, and X. Guan, "Non-cooperative power control for wireless ad hoc networks with repeated games," *IEEE J. Sel. Areas Commun.*, vol. 25, no. 6, pp. 1101–1112, Aug. 2007.

22. M. Bowling and M. Veloso, "Rational and convergent learning in stochastic games," in *Proc. Int. Joint Conf. Artificial Intelligence*, pp. 1021–1026, Seattle, WA, Aug. 2001.

23. S. J. Pan and Q. Yang, "A survey on transfer learning," *IEEE Trans. Knowl. Data Eng.*, vol. 22, no. 10, pp. 1345–1359, Oct. 2010.

24. L. Xiao, T. Chen, C. Xie, H. Dai, and H. V. Poor, "Mobile crowdsensing games in vehicular networks," *IEEE Trans. Vehicular Technology*, Jan. 2017.

Chapter 6
Conclusion and Future Work

In this book, we first propose a PHY-layer authentication scheme based on ambient radio signals and the RSSI of packets that usually ignored in VANETs. The problem of network selection in VANETs is considered, taking into account the rapid changes in signal strength brought about by high-speed movement of the vehicle. In addition, we propose a hotbooting PHC-based UAV relay strategy to resist smart jamming without the knowledge of the UAV channel model and the jamming model. A learning-based task offloading framework using the multi-armed bandit theory is developed, which enables vehicles to learn the potential task offloading performance of its neighboring vehicles with excessive computing resources and minimizes the average offloading delay [1].

In this section, we summarize the future learning based VANET communications in Sect. 6.1, and list several learning based security techniques in VANETs in Sect. 6.2. Next-generation VANETs will have special requirements of autonomous vehicles with high mobility, low latency, real-time applications, and connectivity, which may not be resolved by conventional cloud computing. Hence, merging of fog computing with the conventional cloud for VANETs is a potential solution for several issues in current and future VANETs. Thus in this book we investigated the task offloading for vehicular cloud computing systems. New learning scheme should be investigated to effectively adapt to the highly dynamic vehicular environment and balance the tradeoff between exploration and exploitation in the learning process with performance guarantees. The proposed learning-based VANET security methods will suffer from a long learning time due to the random exploration at the beginning, the estimation error and delay regarding the network state, and reward in the dynamic game. Thus, backup security solutions in VANETs have to be designed in the future works.

© Springer Nature Switzerland AG 2019
L. Xiao et al., *Learning-based VANET Communication and Security Techniques*, Wireless Networks, https://doi.org/10.1007/978-3-030-01731-6_6

6.1 Future Learning Based VANET Communications and Computing

Vehicles have become the third most important space outside of homes and work-places. By deploying computing and storage resources at the network edge, mobile edge computing (MEC) can provide low latency computing services. However, several issues need to be solved, such as the heterogeneous cloud architecture with vehicular cloud computing (VCC), mobile edge computing and remote cloud, as well as the cooperation of service vehicles (SeVs). New learning scheme should be investigated to effectively adapt to the highly dynamic vehicular environment and the heterogeneous network architecture and balance the tradeoff between exploration and exploitation in the learning process with performance guarantees, such as the average offloading delay.

A key challenge of vehicular edge computing (VEC) is to guarantee the quality of service (QoS) requirements of computation tasks, including offloading delay and service reliability, under the *dynamic* vehicular environment. As we have discussed in Chap. 3, a possible solution to improve the service reliability is to introduce *task replication* technique, i.e., the replicas of each task can be offloaded to multiple vehicles and processed by them simultaneously [2, 3]. And we have proposed a learning-based task replication algorithm in Sect. 3.4, to help task generators learn the complex and dynamic vehicular environments such as channel states and computing capabilities. However, a key problem remained is that, what is the optimal number of task replicas to be offloaded under different traffic conditions. While task replication can improve the service reliability when computing resources are sufficient, too many replicas may cause long service waiting time and transmission collisions, which in turn degrades the QoS. To solve this problem, centralized controllers or vehicles can be aware of density of vehicles, computing resources and task workloads, and dynamically adjust the task replicas to optimize the QoS.

Moreover, computation tasks from different types of applications have different requirements in terms of QoS, computing workloads and data volumes. How to manage the computing and communication resources in the VEC system to satisfy the heterogeneous task requirements is also an open problem. Vehicles and RSUs can also cooperate with each other to process tasks with heavy computing workloads. In this case, each task may be partitioned into several subtasks and offloaded to different computing nodes to be processed cooperatively. Machine learning techniques can be used to help the VEC system learn the optimal resource management algorithm for the offloading of heterogeneous tasks and the task partition schemes.

The predictable mobility of vehicles and properties of tasks can bring opportunities to further enhance the QoS of task offloading. To be specific, based on the big data analysis, the routes of vehicles and their instantaneous speeds can be predicted, and the future computation and communication requirements of tasks are foreseeable. Therefore, we can exploit these predicted information to preserve the computation and communication resources for tasks in advance, especially for the tasks with stringent delay requirements.

An architecture tailored for vehicle communications in order to enable automatic learning of the wireless environment is a promising research topic. Due to the high level of mobility and dynamic communication environment, new learning architecture to achieve the intelligence is needed for the adaptive channel selection problem in dynamic spectrum access of vehicle communications.

6.2 Future Learning Based VANET Security

Existing learning-based security techniques involve jamming resistance, PHY-layer authentication and access control will suffer from a long learning time and their performance will degrade in VANETs, due to the random exploration at the beginning, the estimation error and delay regarding the network state, and reward in the dynamic game. For instance, the anti-jamming UAV relay transmission scheme as presented in [4] uses UAV to relay the OBU messages can not meet the jamming resistance performance due to the SINR and BER estimation error and transmission delay of the backup RSU.

Therefore, several challenges have to be addressed to implement the learning based security schemes in practical VANET systems:

The mobile device and the UAV usually have difficulty estimating the current network and attack state accurately and fast enough to choose the next defense policy. Therefore, we have to investigate the impacts of the inaccurate and delayed state information on the VANET security performance. In addition, we have to improve the learning-based VANET security solutions with the advanced learning techniques that require less state information and tolerate the inaccurate and delayed state observation for VANET communication systems. A promising solution is to incorporate the known network and attack information extracted with transfer learning and data mining techniques to accelerate the learning process.

The mobile device and the UAV has to observe the security gain and the protection cost to evaluate its reward from each action. Both in turn consists of a large number of factors. For example, in a secure mobile offloading [5], the mobile device has to accurately evaluate the data importance, the transmission and computation delay, the energy cost and the rogue edge risks from its last offloading policy, and incorporates them properly to evaluate the utility, which is challenging for most practical VANET systems. The VANET communication systems have to investigate these factors in the utility evaluation instead of using the heuristic model used in most existing learning-based security schemes. It is critical to replace the heuristic RL methods such as Q-learning in the VANET security solutions with the newly developed RL techniques such as deep learning that work well with delayed and inaccurate utility information.

The learning-based security methods have to explore the "bad" security policy that sometimes can cause network disaster for VANET systems at the beginning stage of learning to achieve the optimal strategy. The rogue edge detection scheme based on Q-learning techniques sometimes have miss detection rates and false alarm rates that are not negligible for VANET systems. These explorations that are dangerous for VANET security indicates a large number of failed defense against attackers. To this end, transfer learning techniques that use data mining to explore existing defense experiences can be designed to help the RL reduce the random exploration and thus decreases the risks of trying bad defense policies at the beginning of the learning process. Backup security solutions have to be designed and incorporated with the learning-based security schemes to avoid a security disaster from a bad decision made in the learning process such as connecting with a rogue edge.

References

1. Y. Sun, X. Guo, S. Zhou, Z. Jiang, X. Liu, and Z. Niu, "Learning-based task offloading for vehicular cloud computing systems," in *Proc. IEEE Int'l Conf. Commun. (ICC)*, Kansas City, MO, USA, May 2018.
   CoRR, vol. abs/1804.00785, Apr. 2018. [Online]. Available: https://arxiv.org/abs/1804.00785.
2. S. Z. X. G. Y. Sun, J. Song and Z. Niu, "Task replication for vehicular edge computing: A combinatorial multi-armed bandit based approach," in *Proc. IEEE Global Commun. Conf. (GLOBECOM)*, Abu Dhabi, UAE, Dec. 2018.
3. X. G. Z. Jiang, S. Zhou and Z. Niu, "Task replication for deadline-constrained vehicular cloud computing: Optimal policy, performance analysis and implications on road traffic," *IEEE Internet Things J.*, vol. 5, no. 1, pp. pp. 93–107, Feb. 2018.
4. L. Xiao, X. Lu, D. Xu, Y. Tang, L. Wang, and W. Zhuang, "UAV relay in VANETs against smart jamming with reinforcement learning," *IEEE Trans. Vehicular Technology*, vol. 67, no. 5, pp. 4087–4097, May 2018.
5. X. Lu, X. Wan, L. Xiao, Y. Tang, and W. Zhuang, "Learning-based rogue edge detection in vanets with ambient radio signals," in *Proc. IEEE Int'l Conf. Commun. (ICC)*, Kansas City, MO, May 2018.

Printed in the United States
By Bookmasters